PHYSICAL METHODS IN ORGANIC CHEMISTRY

B. I. Ionin and B. A. Ershov
NMR Spectroscopy in Organic Chemistry, 1970

V. I. Minkin, O. A. Osipov, and Yu. A. Zhdanov
Dipole Moments in Organic Chemistry, 1970

DIPOLE MOMENTS
IN ORGANIC CHEMISTRY

DIPOLE MOMENTS
IN ORGANIC CHEMISTRY

Vladimir I. Minkin, Osip A. Osipov, and Yurii A. Zhdanov

Department of Chemistry
Rostov University
Rostov-on-Don, USSR

Translated from Russian by
B. J. Hazzard

Translation edited by Worth E. Vaughan
Department of Chemistry
University of Wisconsin
Madison, Wisconsin

℗ PLENUM PRESS • NEW YORK–LONDON • 1970

547.1377
M665

Library of Congress Catalog Card Number 69-17901

SBN 306-30408-2

The original Russian text was first published by Khimiya Press in Leningrad in 1968. The present translation is published under an agreement with Mezhdunarodnaya Kniga, the Soviet book export agency.

Владимир Исаакович Минкин
Осип Александрович Осипов
Юрий Андреевич Жданов

ДИПОЛЬНЫЕ МОМЕНТЫ

DIPOL'NYE MOMENTY

© 1970 Plenum Press, New York
A Division of Plenum Publishing Corporation
227 West 17th Street, New York, N. Y. 10011

United Kingdom edition published by Plenum Press, London
A Division of Plenum Publishing Corporation, Ltd.
Donington House, 30 Norfolk Street, London W. C. 2, England

Preface

In accordance with the aims of the series "Physical Methods in Organic Chemistry," of which this book forms part, the authors' main aim was a systematic account of the most important methods of using the method of dipole moments in organic chemistry and interpreting its results in practice.

Since 1955, when two monographs devoted to the fundamentals and applications of the dipole moment method appeared simultaneously (C. P. Smyth, Dielectric Behavior and Structure, McGraw-Hill, New York; and J. W. Smith, Electric Dipole Moments, Butterworths, London), no generalizing studies of this type have appeared in the Russian and foreign literature. Nevertheless, it is just in this period that almost half of all publications on the structure and properties of organic compounds by means of the dipole moment method have appeared.

During this time, the principles of the method of measurement and the physical theory of the method have not undergone fundamental changes. Consequently, in giving an account of these matters we considered it sufficient to give a very short introduction to the theory of the method that is not burdened with details of the mathematical derivations and the strict formalism of the theory of dielectrics which are hardly used in the applications of the method that are of interest to the organic chemist (Chapter I).

In Chapter II the experimental methods of determining dipole moments are discussed in detail. Here the main attention has been devoted to the method of determining dipole moments in solutions in nonpolar solvents, which is the method most widely used for studying organic compounds. Here again, intermediate stages in

v

the calculations are not given and all the formulas necessary for
treating the results of measurement are given in a form directly
suitable for carrying out the calculations.

In none of the published handbooks on dipole moments has
the numerical apparatus of the method which is necessary for the
structural interpretation of the results of the determinations been
given systematically. In our opinion, it is partly for this reason
that it is frequently possible to come across the idea of the exces-
sively limited nature of the method in the study of fine features of
the steric and electronic structure of organic compounds. The
material collected in Chapter III should fill in this gap to a certain
extent and facilitate the analysis of the dipole moments of organic
compounds.

Chapters IV and V consider the fields of the traditional application of the method of dipole moments in organic chemistry:
investigations of the spatial and electronic structure of molecules.
In contrast to preceding monographs and reviews on dipole moments the material is arranged systematically not according to the
type of organic compounds but according to the nature of the structural problems: conformational analysis, internal rotation, various types of electronic effects, and so on. Such an approach to a
consideration of the applications of the method, it appears to us,
corresponds more accurately to the spirit of modern physical organic chemistry.

The possibilities presented by the dipole moment method for
studying some specific problems of the structure of organic compounds (tautomerism, the hydrogen bond, and other types of intermolecular interactions) are considered in Chapter VI. A special
position is occupied by a section on the dipole moments of organic
compounds in electronically excited states.

The scope and purpose of the monograph have not permitted
adequate attention to be devoted to a whole series of questions having direct relationship with the method of dipole moments. This
relates in the first place to the principles of the construction of
the measuring apparatus. Readers with a special interest in this
field we must refer to the books by Smyth and Smith already mentioned and also to the extremely detailed review in the third volume of the series "Technique of Organic Chemistry" (A. Weissberger, ed., Wiley, New York). However, we hope that we have

succeeded in selecting those branches that are of the greatest interest for the organic chemist in his current work.

Chapters I and II were written by O. A. Osipov, Chapter III by V. I. Minkin, Chapters IV and V by V. I. Minkin and Yu. A. Zhdanov, and Chapter VI by V. I. Minkin and O. A. Osipov.

We shall be sincerely grateful to receive any information concerning deficiencies in the book.

CHAPTER VI

Dipole Moments and Some Special Problems

of the Structure and Properties

of Organic Compounds

Chapter I

Basic Principles of the Theory of Dielectrics

1. Behavior of a Dielectric in a Static Electric Field*

An investigation of the electrical characteristics of a molecule gives important information on the distribution of charges in the molecule and provides the possibility of determining many properties of the molecule which depend on its electronic distribution. Those electrical properties of the molecule must be selected that are capable of a theoretical interpretation. The classical theory of the polarization of dielectrics shows that such properties of a molecule are exhibited in the behavior of the substance in an electric field.

Consider the behavior of a dielectric in a static electric field. Let us imagine a condenser with plane-parallel plates separated from one another by the distance r which is small in comparison with their linear dimensions. If the plates are charged and the surface density of the charges on them is $+\sigma$ and $-\sigma$, a practically uniform field is created in the condenser in a direction perpendicular to the surfaces of the plates. The strength of this field in vacuum will be

$$E_0 = 4\pi\,\sigma \tag{I.1}$$

---

* The theory of polarization is discussed in more detail in References [1-12].

1

Under these conditions, the difference in potential V arising between the plates of the condenser can be defined as

$$V = |E_0|r \qquad (I.2)$$

Let us fill the space between the plates of the condenser with a dielectric, keeping the charge density on the plates of the condenser at its previous value. This leads to a fall in the potential difference between the plates of the condenser by the amount V/ε_s, where ε_s is the static dielectric constant of the substance. Since relation (I.2) remains valid, the strength of the electric field decreases by the same magnitude

$$E = \frac{4\pi\sigma}{\varepsilon_s} \qquad (I.3)$$

Thus, the static dielectric constant may be considered as the ratio of the field strength in vacuum to the field strength of the condenser containing the dielectric:

$$\varepsilon_s = \frac{E_0}{E} \qquad (I.4)$$

The static dielectric constant is easily expressed in terms of the capacitances of the condenser in vacuum and when it is filled with dielectric, since the capacitance is q/V. Here q is the charge density, σ, times the area of the condenser plates. Then we obtain

$$\varepsilon_s = \frac{C}{C_0} \qquad (I.5)$$

where C_0 and C are, respectively, the capacitance of the condenser in vacuum and its capacitance when filled with the dielectric.

The dielectric constant is generally determined by measuring the capacitances C_0 and C [3, 5, 7, 8, 12, 13] and then using formula (I.5), except for those cases where measurements are carried out at very low or very high frequencies, for which other methods are used [5, 13].

From a macroscopic standpoint, the influence of an electric field on a dielectric, leading to an increase in the capacitance of the condenser, is equivalent to the charging of the two surfaces of the dielectric directly adjacent to the plates of the condenser with charges of opposite signs (Fig. 1).

Such an accumulation of uncompensated negative charges on the surface of the dielectric adjacent to the positively charged plate and of positive charges adjacent to the negatively charged plate of the condenser leads to a partial decrease in the original

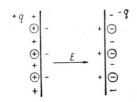

Fig. 1. Macroscopic description of the change in the potential difference on introduction of a dielectric between the plates of a plane-parallel condenser.

charges q. This follows from equations (I.1) and (I.3), a combination of which leads to the expression

$$P = \sigma \frac{\varepsilon_s - 1}{\varepsilon_s} \qquad (\text{I.6})$$

The magnitude P must be considered as the surface charge density on the dielectric.

The cause of the increase in the capacitance of the condenser is the polarization of the dielectric under the action of the applied electric field.

In the absence of an electric field, the substance as a whole is electrically neutral and in any small volume of it (which must, however, contain a sufficiently large number of molecules) the centers of all the positive and negative charges coincide. Under the action of the applied field, a displacement of the centers of gravity of the charges by some distance $\vec{l}$ takes place and an electric dipole appears.* Such a displacement of the charges under the action of a field is called the e l e c t r i c   p o l a r i z a t i o n $\vec{p}$ of the substance. The phenomenon of polarization can be considered by ascribing to each small volume of the dielectric some induced dipole moment. This is valid since the electric dipole moment can characterize not only the electric state of the individual molecule† but also that of some macroscopic volume of the dielectric consisting of a large number of molecules. Then, for unit volume of the dielectric (1 cm³) the magnitude of the induced dipole moment can be given as

$$\vec{p} = \sum_i e_i \vec{l}_i \qquad (\text{I.7})$$

---

* In the general case, by an electric dipole must be understood any system consisting of electric charges q equal in magnitude and opposite in sign separated by a distance vector $\vec{l}$. The magnitude of such a dipole is defined by its electric moment:

$$\vec{\mu} = q\vec{l}$$

† The behavior of an individual molecule in an electric field will be considered below.

where $\vec{p}$ is the polarization per unit volume. Here the summation is extended over all the charges (electrons and nuclei) present in unit volume of the dielectric.

It follows from what has been said above that the field within the dielectric $\vec{E}$ must be composed of the field $\vec{E}_0$, created by the charges q on the plates of the condenser when the dielectric substance is absent and the field induced by the dipoles, which is in the opposite direction to $\vec{E}_0$.

According to the laws of electrostatics, the field created by the induced dipoles is $-4\pi\vec{p}$. Consequently, for the field within the dielectric

$$\vec{E} = \vec{E}_0 - 4\pi\vec{p} \tag{I.8}$$

where $\vec{E}$ is due both to the charge density on the plates of the condenser and to the charge density on the surface of the dielectric.

In the macroscopic theory of dielectrics, the vector $\vec{D}$, which is called the electric displacement or the electric induction, is introduced:

$$\vec{D} = \vec{E} + 4\pi\vec{p} \tag{I.9}$$

The connection between the electric induction and the field strength within a dielectric can also be determined on the basis of equation (I.4):

$$\vec{D} = \vec{E}\varepsilon_s \tag{I.10}$$

As can be seen from (I.10) the magnitude $\vec{D}$ is proportional to $\vec{E}$, the proportionality factor being the static dielectric constant. The difference between $\vec{D}$ and $\vec{E}$ depends on the degree of polarizability of the dielectric in an electric field. In vacuum, where there is no polarization, $\vec{D} = \vec{E}$ and $\varepsilon_s = 1$.

Thus, in the macroscopic theory, the electric field in dielectrics is described by means of two quantities: the macroscopic field $\vec{E}$ and the electric induction $\vec{D}$.

From equations (I.9) and (I.10), we obtain

$$\vec{p} = \frac{\varepsilon_s - 1}{4\pi} \vec{E} \qquad\qquad (I.11)$$

Formula (I.11) establishes a connection between the dielectric constant, the field $\vec{E}$, and the polarization of the dielectric $\vec{p}$. The polarization vector is proportional to the field and has the same direction. However, this is valid only for isotropic media. In anisotropic media the direction of the polarization vector may or may not coincide with the direction of the field. In this case, the absolute magnitude of the vector $\vec{p}$ depends not only on the absolute value of the vector $\vec{E}$, but also on its direction with respect to the principal axes of the dielectric.

The further consideration of the polarization of a dielectric $\vec{p}$ requires the use of molecular ideas. In order to connect the macroscopic behavior of a dielectric with the properties of its individual molecules and to establish the polarization mechanism it is necessary to determine how an isolated neutral molecule of the dielectric will behave in an electric field. From this point of view, all dielectrics may be divided into nonpolar and polar media. In the former, the molecules possess electrical symmetry and the centers of gravity of the positive and of the negative charges coincide. Polar dielectrics, on the other hand, are constructed of electrically asymmetrical molecules in which the centers of gravity of the positive and of the negative charges are located at some distance l from one another and form an electric dipole. Thus, nonpolar molecules do not possess a dipole moment in the absence of a field, while polar molecules have a permanent dipole moment independent of the field.

Let us first consider the behavior of molecules possessing no permanent dipole moments.

Under the action of an applied electric field, a process of polarization takes place not only in any small volume of the dielectric but also in each individual molecule, whether or not it has a permanent dipole moment in the absence of a field. The action of the field leads to the appearance in the molecule of some induced dipole moment $\vec{m}$, the magnitude of which is proportional to the strength of the mean macroscopic field $\vec{E}$

$$\vec{m} = \alpha\vec{E} \tag{I.12}$$

where α is a proportionality factor, the so-called polarizability of the molecule. It is a measure of the mobility of the charges in the molecule and characterizes their relative displacement under the action of the field.

If unit volume (1 cm^3) contains n molecules possessing the induced dipole moment $\vec{m}$, the polarization $\vec{p}$ will be

$$\vec{p} = n\vec{m} \tag{I.13}$$

Using equations (I.11), (I.12), and (I.13), it is possible to establish the connection between the static dielectric constant, the polarizability of the molecule, and the strength of the mean macroscopic field:

$$\vec{p} = n\vec{m} = n\alpha\vec{E} = \frac{\varepsilon_s - 1}{4\pi}\vec{E} \tag{I.14}$$

whence

$$\varepsilon_s = 1 + 4\pi n\alpha$$

It follows from the last expression that the magnitude of the static dielectric constant is the greater the greater the density of the substance and the greater the polarizability of the molecule.

Equation (I.14) is not strictly correct and has a limited application. To a first approximation, it can be applied to gases at sufficiently low pressures when the distance between the molecules is so large that it is possible to neglect the action of the electrostatic field of the surrounding molecules on the particular molecule with which we are dealing. This condition means that the polarization due to the introduction of a gaseous dielectric is extremely slight, so that the dielectric constant differs very little from unity.

When the distance between neighboring molecules is comparable with the dimensions of the molecules themselves (gases at high densities, liquids) it is no longer possible to neglect the action of the electrostatic field of the surrounding molecules on the particular molecule with which we are dealing. In this case, the dielectric medium can no longer be regarded as a continuous medium with the dielectric constant ε_s as was assumed in the macroscopic discussion. Under the action of the external field, the electrostatic field created by the surrounding molecules is distorted, since the

molecules are polarized and may, in their turn, influence the surrounding molecules. As a result, each molecule of the dielectric
exists under the action of some resultant field which we shall call
the local or effective field. Thus, the local field is the resultant of
the macroscopic field $\vec{E}$ and the internal field due to the electrostatic interaction between the molecules. It is clear that in calculating the dielectric constant, this interaction between the molecules must be taken into account. Lorentz was the first to consider
a method of calculating the local field. According to his model

$$\vec{E}_{loc} = \vec{E} + \vec{E}_1 + \vec{E}_2 \qquad (I.15)$$

Here $\vec{E}_1$ is the field created by the molecules of the dielectric present outside a macroscopic sphere cut in the dielectric and $\vec{E}_2$ the
field created by the molecules present within the sphere.

Lorentz showed that

$$\vec{E}_1 = \frac{4}{3}\,\pi\vec{p} \qquad (I.16)$$

In the general case, it is impossible to calculate $\vec{E}_2$. However, it
can be shown that under some conditions $\vec{E}_2 = 0$. In actual fact,
with a random distribution of molecules separated from one another
by a distance considerably exceeding the dimensions of the molecules themselves (gases, nonpolar liquids), for each molecule within the Lorentzian sphere it is always possible to find another that
compensates the action of the former. Then, to a first approximation, it can be considered that $\vec{E}_2 = 0$.

Under these conditions, the local field will be

$$\vec{E}_{loc} = \vec{E} + \frac{4}{3}\,\pi\vec{p} \qquad (I.17)$$

It follows from what has been said that the behavior of the
induced dipole moments in a molecule of dielectric in an electric
field is due to the local field. Then equation (I.12) must be replaced by the following

$$\vec{m} = a\vec{E}_{loc} = a\left(\vec{E} + \frac{4}{3}\,\pi\vec{p}\right) \qquad (I.18)$$

On the basis of equations (I.11), (I.13), and (I.18) it is easy to
obtain the well-known Clausius–Mosotti formula:

$$\frac{\varepsilon_s - 1}{4\pi n}\,\vec{E} = a\left(\vec{E} + \frac{\varepsilon_s - 1}{3}\,\vec{E}\right)$$

whence

$$\frac{\varepsilon_s - 1}{\varepsilon_s + 2} = \frac{4}{3}\pi n\alpha \qquad (I.19)$$

Formula (I.19) establishes a connection between the static dielectric constant and the polarizability of the molecule, on the one hand, and the number of molecules per unit volume, i.e., the density of the substance, on the other hand.

By multiplying both sides of equation (I.19) by the molar volume M/d, we obtain

$$\frac{\varepsilon_s - 1}{\varepsilon_s + 2} \cdot \frac{M}{d} = \frac{4}{3}N\pi\alpha \qquad (I.20)$$

where N is Avogadro's number, M is the molecular weight and d is the density of the substance. The quantity $\frac{4}{3}\pi N\alpha$ is called the molar polarization and is denoted by P. As can be seen from formula (I.20) the molar polarization no longer depends on the density of the substance and the temperature but only on the polarizability of the molecules α.

The Clausius − Mosotti formula is strictly applicable only to nonpolar gases. It is also approximately valid for polar gases at very low pressures when the static dielectric constant differs only very slightly from unity and for nonpolar liquids (if the short-range interaction between the molecules is neglected). It is completely unsuitable for polar liquids when the interaction of the dipolar molecules at close distances creates additional internal fields ($\vec{E_2}$).

2. Molecular Polarizability

Each type of polarizability is characterized by a definite type of displacement of the charges of the particles of the dielectric under the action of the applied electrostatic field. In the general case, all types of polarizability can be reduced to two main types: 1) elastic displacement of the charges in the atoms and molecules under the action of the electric field, and 2) the orientation of the permanent dipoles in the direction of the applied field.

Electronic Polarization. This type of polarizability is characterized above all by the elastic displacement of the electron charge cloud relative to the nuclei when the atom or molecule is acted upon by an electric field. It is customary to call this polarizability the e l e c t r o n i c   polarizability (α_e) and the magnitude referred to 1 mole of the dielectric substance the e l e c t r o n i c   p o l a r i z a t i o n ($P_e = \frac{4}{3} \pi N \alpha_e$).

Electronic polarizability exists in all atoms and molecules of both polar and nonpolar dielectrics, regardless of the possibility of the appearance of other types of polarizability in the dielectric.

The time required for displacement of the charges as a result of the establishment of electronic polarizability is extremely low, of the order of 10^{-14} -10^{-16} sec, which is comparable with the period of luminous vibrations.

Since the electronic polarizability characterizes the perturbation of the electron orbital, its numerical values must be of the same order as the dimensions of the electron charge cloud, i.e., the dimensions of atoms and molecules. Starting from equation (I.12) and the Coulomb law it can easily be shown that the electronic polarizability has the dimensions l^3:

$$\alpha_e = \frac{[e] \, [l]}{[e]} \, [l]^2 = l^3$$

The numerical value of α_e is of the order of 10^{-24}cm^3. Experimental data and quantum-mechanical calculations give results of the same order for the electronic polarizability. For example, in the case of a spherically symmetrical atom a quantum-mechanical calculation leads to the magnitude $\frac{9}{2}r^3$ (where r is the radius of the atom) for α_e.

With an increase in the volume of the electron charge cloud, the electronic polarizability increases in magnitude. The further the electron is from the nucleus, the greater is its mobility and the more highly is it subject to the action of an electric field. The highest polarizability is possessed by the valence electrons, as those most feebly bound to the nucleus.

Thus, with an increase in the main quantum number, the electronic polarizability must rise. An increase in the number of electrons in one and the same orbital must also lead to a rise in the polarizability (α_e) since each of the electrons will respond to the influence of the applied electric field. Generally speaking, with an increase in the number of electrons (with the same main quantum number) the electronic polarizability may either increase or decrease. This depends on which of two effects predominates, the effect of the increase on the number of electrons or the effect of the decrease in the Bohr radii of the electron orbitals. We may note that the electronic polarizability must also depend on the orbital quantum number l. For example, the p-electrons, which possess a greater mobility than the s-electrons, must be more subject to the action of an electric field. For molecules containing no conjugated bonds, the polarizability can be regarded as the algebraic sum of the polarizabilities of the individual atoms or bonds. When conjugated bonds are present in the molecule, the electronic polarizability exceeds the additive value, which is explained by the greater mobility of the π-electrons in a conjugated system.

We have already mentioned that electrons possess such a low inertia that the time of establishment of the electronic polarization in a molecule under the action of an electric field is comparable with the period of luminous vibrations. This provides the possibility of applying to nonpolar dielectrics the molecules of which possess only electronic polarizability in the electric field the relation

$$n^2 = \varepsilon_s \mu' \tag{I.21}$$

which follows from Maxwell's electromagnetic theory of light.* In this equation, n is the refractive index, ε_s is the static dielectric constant, and μ' is the magnetic permeability. For all diamagnetic substances, μ' differs from unity by less than 10^{-5}, and therefore for all organic compounds with the exception of free radicals the value of μ' can be taken as unity. Then

$$n^2 = \varepsilon_s \tag{I.22}$$

The equality (I.22) is valid for the region of wavelengths sufficiently remote from the region of the absorption bands of the molecule. This must be explained by the fact that Maxwell's theory does not take into account the dependence of the refractive index on the wavelength of the light. In actual fact, in considering formula (I.20) we assume that the polarizability is a constant magnitude which does not depend on the strength of the applied field. This is true for the case of electrostatic fields or low-frequency variable fields. If, however, the dielectric is present in a high-frequency variable field (region of visible light), the polarizability is a function of the frequency of the field.

---

* The high-frequency electromagnetic vibrations of visible light cause practically no displacement of nuclei of atoms and molecules and do not orient permanent dipoles.

In the case of the simplest model, considering an electron in a molecule as a harmonic oscillator with a natural frequency of vibration ν_i, the following expression is obtained for the electronic polarizability in a luminous field of frequency ν:

$$\alpha_e = \frac{e^2}{4\pi^2 m} \sum_i \frac{f_i}{\nu_i^2 - \nu^2} \tag{I.23}$$

Here the summation is carried out over all the electrons in the molecule; f_i is the strength of the oscillator characterizing the degree of participation of the electron in the vibration concerned, and e and m are the charge and mass of an electron.

The value for the static electronic polarizability is found similarly

$$\alpha^0 = \frac{e^2}{4\pi^2 m} \sum_i \frac{f_i}{\nu_i^2} \tag{I.24}$$

The following relations can be derived from equations (I.20), (I.22), and (I.24):

$$\frac{n^2 - 1}{n^2 + 2} = \frac{N_1 e^2}{3\pi m} \sum_i \frac{f_i}{\nu_i^2 - \nu^2} \tag{I.25}$$

$$\frac{\varepsilon_s - 1}{\varepsilon_s + 2} = \frac{N_1 e^2}{3\pi m} \sum_i \frac{f_i}{\nu_i^2} \tag{I.26}$$

The difference between the square of the refractive index n^2 and the static dielectric constant can be clearly seen from equations (I.25) and (I.26). Formula (I.25) shows that with a decrease in the frequency ν (an increase in the wavelength), the refractive index falls. For a stricter substantiation of equality (I.22), the refractive index and the dielectric constant must be determined at the same wavelength. For this purpose the values of the refractive index n must be extrapolated to infinite wavelengths and the value of n_∞ at $\lambda = \infty$ must be found. The value of n_∞ (or the molecular refraction R_∞ can be determined by means of both dispersion formulas and graphical extrapolation methods [14]. In particular, it is possible to use Cauchy's formula, which expresses the dependence of the refractive index on the wavelength

$$n = A + \frac{B}{\lambda^2} + \frac{C}{\lambda^4} \tag{I.27}$$

where A, B, and C are empirical constants determined by measuring the refractive index for three wavelengths. The last member of this equation is frequently neglected, and then

$$n = A + \frac{B}{\lambda^2} \qquad (I.28)$$

The simplest method of finding n_∞ from formula (I.28) consists in measuring two values of the refractive index at two different wavelengths (n_{λ_1} and n_{λ_2}). Then formula (I.28) assumes the form

$$n_\infty = \frac{n_1 \lambda_1^2 - n_2 \lambda_2^2}{\lambda_1^2 - \lambda_2^2}$$

Dispersion formulas are not very accurate, the error amounting to several parts per thousand. More reliable and accurate results can be obtained by making use of the graphical extrapolation method. If the function $(n^2 - 1)/(n^2 + 2) = 1/f(n)$ is plotted along the axis of ordinates and the magnitude $1/\lambda^2$ along the axis of abscissas, we obtain almost straight lines that can easily be extrapolated to the axis of ordinates ($\lambda = \infty$). We may note that for the majority of substances absorbing in the ultraviolet region of the spectrum, n_∞ differs only very slightly from the refractive index for the yellow sodium line (5893 Å) (n_D) and therefore for practical purposes the value of the refractive index n_D is frequently used in place of n_∞ for calculating molecular refractions.

By substituting in formula (I.20) the square of the refractive index in place of the static dielectric constant, we obtain the well-known Lorentz – Lorenz* formula describing the optical behavior of a substance:

$$R = \frac{n^2 - 1}{n^2 + 2} \cdot \frac{M}{d} = \frac{4}{3} \pi N \alpha_e \qquad (I.29)$$

where R is the m o l a r   r e f r a c t i o n.

To a first approximation, the molar refraction, like the electronic polarization, does not depend on the temperature. This is explained by the fact that the difference in energies between the normal and excited states of the electrons is very large and the probability of transition of an electron into excited states is very low.

---

* The Lorentz–Lorenz formula can be derived in a completely different manner on the basis of optical phenomena alone without using the theory of the dielectric constant.

The molar refraction may be regarded as a measure of the polarizability of the molecules in the electromagnetic field of visible light due to the elastic displacement of the electron clouds. Consequently, it characterizes the electronic polarizability of the molecule. A considerable amount of numerical material has been accumulated on molar refraction and is frequently used in calculations of dipole moments.

However, it must be borne in mind that the Lorentz-Lorenz formula is a first approximation, since it is based on the simplified ideas relative to the internal field $\vec{E}_2$ that were discussed above.

Up to the present time, in considering the polarizability of a molecule we have not taken into account one extremely important factor, which is that because of the dissimilar dimensions of the electronic clouds of the molecule in different directions, their elastic displacements in an electric field will be different in different directions. Because of this, the electronic polarizability of a molecule in the general case must possess electric anisotropicity in space.

In this case, the polarizability is described by a tensor which can be represented in the form of an ellipsoid of polarization. The property of such an ellipsoid consists in the fact that the polarizability of the molecule in space can be reduced to its polarizability in three mutually perpendicular directions α_1, α_2, and α_3, corresponding to the three principal semiaxes of the ellipsoid. If an electric field of unit strength is applied along each of these three directions, the lengths of the semiaxes will correspond to the values α_1, α_2, and α_3. If the applied electric field of unit strength is directed at an arbitrary angle to the three main semiaxes of the ellipsoid, it is possible to resolve the field into components relative to the three main directions and to determine the polarizability for each of them separately. The total value of the polarizability is obtained by vector summation.

For isotropic molecules (for example, CCl_4, SF_6) with a spherically symmetrical spatial distribution of the electron cloud, all three components of the polarizability are equal:

$$\alpha = \alpha_1 = \alpha_2 = \alpha_3 \qquad (I.30)$$

In anisotropic molecules, the polarizability along the direction of the valence bonds is always greater than in the other directions. This is well illustrated by the data given in Table 30.

Various methods (the use of the Kerr effect, the measurement of the depolarization of scattered light) are known which enable us to find the total polarizability of a molecule α_e through its individual components and to determine the polarizability of bonds.

Atomic Polarization. In an electric field not only the electron clouds but also the actual nuclei of the atoms forming the molecule are affected, being displaced relative to one another. This form of polarizability of the molecule is usually called the atomic polarizability α_a and the polarization corresponding to it the atomic polarization P_a.

The time to establish the atomic polarizability coincides with the period of the vibrations in the infrared region of the spectrum. This is explained by the fact that the vibrations of the atomic nuclei, because of their low frequency, cannot be excited by visible light and are excited only by infrared light. Consequently, the atomic polarization can be determined in the infrared region of the spectrum.

One of the methods for determining the atomic polarization P_a is measuring the refractive index at infrared frequencies and then calculating the results by means of the dispersion formula

$$P_a = \frac{N}{9\pi} \sum_i \frac{e_i^2}{m_i \left(v_i^2 - v_0^2 \right)} \qquad (I.31)$$

where e_i and m_i are the effective charge and mass associated with the corresponding vibrations of the molecule of frequency v_i and v_0 is the frequency of the incident radiation.

In view of the fact that the dielectric constant is determined in a low-frequency electric field, equation (I.31) can be given in the form

$$P_a = \frac{N}{9\pi} \sum_i \frac{e_i^2}{m_i v_i^2} \qquad (I.32)$$

By means of equation (I.32) it is possible to calculate the value of the atomic polarization if the reduced mass and frequency

of vibrations of the molecule are known, the latter being determined from the infrared absorption spectrum. We may note that
the effective charge appearing in equation (I.32) is not the charge
of the nuclei and the electrons since this does not characterize the
charge of the individual particle. It can be regarded as the derivative $d\mu/dr$ of the dipole moment with respect to the distance r between the nuclei. The effective charge of molecules of nitrogen,
hydrogen, and the like is zero and, therefore, regardless of the
frequency of the vibrations, these molecules must have zero atomic
polarization.

The determination of the atomic polarization by means of
equations (I.31) and (I.32) involves considerable difficulties, and
therefore it is found by an indirect method* (see Chapter II).

The atomic polarization is comparatively small, being considerably less than the electronic polarization. In actual fact, both
experimental results and some theoretical conclusions show that
the atomic polarization amounts to a small fraction, not more than
10%, of the electronic polarization. In calculating dipole moments
the atomic polarization is generally assumed to be 5-10% of the
electronic polarization.

It follows from what has been said that the sum of the electronic and atomic polarizations $P_e + P_a$ is the deformation polarization P_d, which characterizes the elastic displacement both of the
electron clouds and of the nuclei of the atoms in the molecule under the action of an electric field.

Consequently, formula (I.20) can be given in the following
form:

$$P_e + P_a = \frac{\varepsilon_s - 1}{\varepsilon_s + 2} \cdot \frac{M}{d} = \frac{4}{3} \pi N \alpha_e + \frac{4}{3} \pi N \alpha_a \qquad (I.33)$$

where $P_e = \frac{4}{3} \pi N \alpha_e$ and $P_a = \frac{4}{3} \pi N \alpha_a$.

Orientation Polarization. The behavior in an electric field of a polar dielectric the molecules of which possess a
permanent dipole moment μ_0 differs from the behavior of a nonpolar dielectric. While for the molecules of the latter the deformation polarizability is the only effect of the action of the applied
field, in the case of the polar molecules in addition to the deforma-

---

* The atomic polarizations of some compounds are given in a handbook [48].

tion polarizability there is another form of polarizability caused
by the orientation of the permanent dipoles in the electric field.
This form of polarizability is called orientation polarizability (α_{or}).
Assuming that the forms of polarizability of a molecule in an electric field considered above are independent of one another, the total polarizability of a molecule in an applied field can be given as
the sum

$$a = \alpha_e + \alpha_a + \alpha_{or} \qquad (I.34)$$

Let us consider briefly the reasons for the appearance of
orientation polarizability and its connection with the permanent dipole moment of the molecule.

In the absence of an electric field, a free molecule with the
moment μ_0 possesses not only translational but also rotational and
vibrational motions. Because of the random motion (distribution),
the mean dipole moment of all the molecules within a small physical volume will be zero. The application of an electric field leads
to the orientation of the dipoles in the direction of the field and to
the appearance in the polar dielectric of some nonzero moment.
This resultant moment will be the higher, under otherwise identical conditions, the higher the strength of the field and the higher
the permanent dipole moment of the molecule.

In an electric field, the molecules of a polar substance will
strive to take up such a position that their dipoles are arranged

parallel to the direction of the field $\vec{E}$. Such an orientation of the
dipoles corresponds to the minimum potential energy. Since the
axis of the dipoles will normally make a nonzero angle (ϑ) with the
direction of the field, even on very high field strengths of the order
of 10^5 V/cm, we obtain for the potential energy U

$$U = - \mu_0 E \cos \vartheta \qquad (I.35)$$

It follows from this equation that the potential energy is the
smaller the smaller the angle ϑ and is a minimum at $\vartheta = 0$.

In a static electric field E, a rotational force which tends to
turn the axis of the dipole in the direction of the field acts on a
molecule with a dipole moment μ_0. The rotational moment T of
this force is

$$T = \mu_0 E \sin \vartheta \qquad (I.36)$$

This force is opposed by the rotational thermal motion of the molecules, which tends to restore random distribution. As a result, a statistical equilibrium is set up. Since the rotational movement of the molecules is a function of the temperature, the orientation polarization must depend on the temperature and is smaller the higher the temperature.

The establishment of orientation polarization when an electric field is applied to a polar dielectric and the process of its disappearance when the field is removed take place considerably more slowly than the establishment of deformation polarization. We may also note that the time to establish orientation polarization depends on a number of factors, in particular, the dimensions of the molecule and the viscosity of the medium.

Debye, using the statistical theory of orientation first developed by Langevin [15] for the permanent magnetic moments of paramagnetic bodies, laid the basis of the theory of the behavior of polar dielectrics in an electric field. Let us consider briefly the main principles of this theory. We shall bear in mind the fact that the following approximations are permitted in a simplified calculation: the field has no perturbing effect on the magnitude of the dipole moment μ_0 of the molecule; the dipoles may assume any position relative to the direction of the applied field; and, finally, the energy of dipole—dipole interaction is insignificant in comparison with the energy of thermal motion (i.e., the density of the gas is very small).

In order to solve the problems we must first establish the connection between the permanent dipole moment of the molecule μ_0 and its orientation polarizability α_{or}, and for this the mean electric dipole moment of the molecule in the direction of the applied field must be determined. Debye approached the solution of this question in the following way.

According to Boltzmann statistics, the number of molecules N_1 included in the solid angle $d\Omega$ will be

$$dN_1 = A_1 e^{-\frac{U}{kT}} \, d\Omega \qquad (I.37)$$

or, taking equation (I.35) into account

$$dN_1 = A_1 e^{\frac{\mu_0 E \cos \vartheta}{kT}} \, d\Omega$$

where A_1 is a proportionality factor which depends on the total number of molecules, k is Boltzmann's constant, and T is the absolute temperature. Each of the dipoles included in the solid angle $d\Omega$ creates a moment equal to $\mu_0 \cos \vartheta$ in the direction of the field. Then the component of the dipole moment of all the molecules in the angle $d\Omega$ in the direction of the field may be given by the following expression:

$$d\mu = \mu_0 \cos \vartheta \, dN_1 = A_1 \mu_0 \cos \vartheta e^{\frac{\mu_0 E \cos \vartheta}{kT}} d\Omega \tag{I.38}$$

The mean moment $\overline{m}$ of a molecule in the direction of the field is expressed by the relation

$$\overline{m} = \frac{\int d\mu}{\int dN_1} = \frac{\int A_1 \mu_0 \cos \vartheta e^{\frac{\mu_0 E \cos \vartheta}{kT}} d\Omega}{\int A_1 e^{\frac{\mu_0 E \cos \vartheta}{kT}} d\Omega} \tag{I.39}$$

In order to perform integration with respect to the angle ϑ, we introduce the following symbols:

$$\frac{\mu_0 E}{kT} = x \tag{I.40}$$

$$\cos \vartheta = y \tag{I.41}$$

After integration over all directions, we obtain

$$\frac{\overline{m}}{\mu_0} = \coth x - \frac{1}{x} = L(x) \tag{I.42}$$

According to (I.40) and (I.42), we have

$$\frac{\overline{m}}{\mu_0} = L\left(\frac{\mu_0 E}{kT}\right) \tag{I.43}$$

Here, $L(\mu_0 E/kT)$ is the Langevin function. It characterizes the ratio of the mean moment of the molecule to its permanent moment as a function of the field strength and the temperature. If $\mu_0 E \ll kT$, the Langevin function can be expanded in a rapidly convergent series, and then equation (I.43) may be limited to the first few members and be given in the form

$$\frac{\overline{m}}{\mu_0} = \frac{\mu_0 E}{3kT}\left(1 - \frac{\mu_0^2 E^2}{15k^2 T^2} + \ldots\right) \tag{I.44}$$

Under the condition $\mu_0 E/kT \ll 1$, we obtain

$$\overline{m} = \frac{\mu_0^2}{3kT} E \tag{I.45}$$

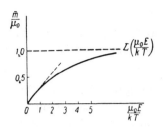

Fig. 2. $\bar{m}/\mu_0$ as a function of $\mu_0 E/kT$.

In equation (I.45) only the effect due to the orientation of the dipoles in the direction of the applied field is reflected. When deformation polarizability is taken into account, we obtain the value for the mean moment in the form

$$\bar{m} = \left(\alpha_d + \frac{\mu_0^2}{3kT}\right)E \qquad (I.46)$$

It follows from equation (I.46) that the effect of the orientation of the dipoles in the electric field is characterized by the term $\mu_0^2/3kT$. Consequently, the orientation polarization can be represented by the magnitude

$$\alpha_{or} = \frac{\mu_0^2}{3kT} \qquad (I.47)$$

Let us now consider the dependence of $\bar{m}/\mu_0$ on $\mu_0 E/kT$, which is shown in Fig. 2. At small values of $\mu_0 E/kT$, i.e., at high temperatures and low field strengths, the segment of the curve approximates to a straight line.

Simple calculation shows that at T = 300°K and with a field strength of 10^5 V/cm, the magnitude of $\mu_0 E$ is hundreds of times smaller than kT (of the order of 10^{-2} -10^{-3}). Consequently, the ratio $\bar{m}/\mu_0$ is extremely small and the orientation effect of the dipoles in the direction of the field is insignificant.

At low temperatures and high fields, the curve $L(\mu_0 E/kT)$ asymptotically approaches the straight line corresponding to the equation $\bar{m}/\mu_0 = 1$, and then the orientation polarization reaches saturation and cannot increase when the field strength is raised further. In actual fact, at $\bar{m}/\mu_0 = 1$, all the dipoles must be oriented in the direction of the field (cos ϑ = 1) and the polarization must have reached its limiting value.

Elementary calculation makes it possible to evaluate the order of the orienting polarizability:

$$\alpha_{or} = \frac{\mu_0^2}{3kT} = \frac{10^{-36}}{3 \cdot 1.38 \cdot 10^{-16} \cdot 3 \cdot 10^2} \approx 10^{-22} cm^3$$

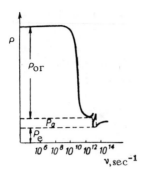

Fig. 3. The polarization P as a function of the frequency ν.

Thus, at T = 293°K for nitrobenzene ($\mu_0 = 3.98$ D), the orientation polarizability is $1.3 \cdot 10^{-22}$ cm^3.

In spite of the fact that under the conditions considered above the orientation polarization is very small, it is still far larger (by one to two orders of magnitude) than the deformation polarization. Consequently, the behavior of polar molecules in an electric field depends mainly on the orientation polarization.

Thus, the overall expression for the polarizability of the molecule is

$$\alpha = \alpha_e + \alpha_a + \frac{\mu_0^2}{3kT} \qquad (I.48)$$

On substituting the value (I.48) in the Clausius−Mosotti formula, we obtain Debye's formula describing the dependence of the dielectric constant of polar dielectrics not only on the distortion polarizability of the molecule but also on the magnitude of the permanent dipole moment and its orientation in the electric field:

$$P = \frac{\varepsilon - 1}{\varepsilon + 2} \cdot \frac{M}{d} = \frac{4}{3} \pi N \left(\alpha_e + \alpha_a + \frac{\mu_0^2}{3kT} \right) \qquad (I.49)$$

where P is the total polarization, equal to the sum of the atomic, electronic, and orientation polarizations ($P = P_e + P_a + P_{or}$). These three components of the polarization are equal, respectively, to:

$$P_e = \frac{4}{3} \pi N \alpha_e; \qquad P_a = \frac{4}{3} \pi N \alpha_a; \qquad P_{or} = \frac{4}{3} \pi N \frac{\mu_0^2}{3kT}$$

The three components of the molecular polarization of polar dielectrics that have been discussed are connected directly with the frequency of the applied field. Figure 3 shows the dependence of P on the frequency of the field over a wide range (from radio waves to ultraviolet waves).

The electronic polarization is inertialess, since the elastic displacement of the electron clouds takes place so fast that it does not lag behind the oscillations of the applied field even at the high frequencies in the region of ultraviolet waves. The atomic polarization is established considerably slower because of the large mass of the atomic nuclei.

The process of establishing the orientation polarization takes place even more slowly and corresponds to the period of radio waves. Consequently, the orientation polarization is one of the so-called slow relaxation types of polarization.

The Debye formula (I.49) includes all the limitations inherent in the Clausius −Mosotti equation and it is therefore approximate and can be applied to polar gases at very low pressures and also (in rough approximation) to extremely dilute solutions of polar substances in a nonpolar solvent. In considering the theory of polar dielectrics, Debye also took no account of the reactive (internal) field of the molecules.

It was mentioned above that for nonpolar dielectrics equations (I.20) and (I.29) are equivalent. In the case of polar dielectrics, however, the inequality

$$\frac{\varepsilon - 1}{\varepsilon + 2} \cdot \frac{M}{d} > \frac{n^2 - 1}{n^2 + 2} \cdot \frac{M}{d}$$

is valid, which shows that the polarizability of the molecule is not exhausted by the elastic displacement of the electron clouds alone. As we have already stated, the cause of this is the orientation of the molecules. However, the Lorentz −Lorenz formula is applicable both to nonpolar and polar molecules, so that polar and nonpolar molecules behave similarly in a lightwave field. Consequently, the Lorentz −Lorenz formula can be used to the same approximation as for molecules of nonpolar dielectrics for the determination of the electronic polarization of polar dielectrics. Then formula (I.49) can be given in the form:

$$P = \frac{\varepsilon - 1}{\varepsilon + 2} \cdot \frac{M}{d} = R_D + \frac{4}{3}\pi N\alpha_a + \frac{4}{3}\pi N \frac{\mu_0^2}{3kT} \qquad (I.50)$$

3. Statistical Theory of the Polarization of Polar Liquid Dielectrics

As we have already mentioned, the first attempt to create a statistical theory of polar dielectrics is due to Debye. He proposed two theories, the first of which was soon disproved by experimental results, while the second was subjected to serious and well-founded criticism in papers by Ansel'm [16]. The latter considered that the assumption made by Debye of the isotropicity of the internal field created by the dipolar molecules when the external field was applied was unsubstantiated and must lead to serious errors.

It is scarcely necessary to dwell on Debye's theories, since their inapplicability to polar liquid dielectrics and even to moderate concentrations of the latter in nonpolar liquids had already become obvious after the appearance of Onsager's well-known papers [17]. Thus, neither of Debye's two theories could in fact establish a relationship between the dielectric constant of a polar liquid and the dipole moment of its molecules. Onsager attempted to solve this problem on the basis of certain model ideas.

Onsager's theory is based on the assumption that a molecule in a polar liquid can be regarded as a pair of poles in the center of which there is a point dipole with a moment equal to

$$\mu_{sum} = \mu_0 + \alpha E_{loc} \tag{I.51}$$

Here μ_{sum} is the total moment. It is composed of the permanent dipole moment μ_0 of the molecule of a polar dielectric and the deformation moment created in the so-called (effective) field (αE_{loc}).

The essence of Onsager's theory is that the local (effective) field is separated into two components: the cavity field G acting in the absence of the total dipole moment μ_{sum}, and the reaction field R, due to the polarization of the medium arising on the introduction of a dipole with moment μ_{sum} into the center of a hollow sphere of radius a. Onsager called the field R the reaction field since it is due to the action of the given molecule on itself through a surrounding sphere with the macroscopic dielectric constant ε_s. Then the local field can be represented as

$$E_{loc} = G + R \tag{I.52}$$

Theoretical calculations (using the Laplace equation) give the following expressions for the cavity field G and the reaction field R:

$$G = \frac{3\varepsilon}{2\varepsilon + 1} R \tag{I.53}$$

$$R = \frac{2(\varepsilon - 1)}{2\varepsilon + 1} \cdot \frac{\mu_{sum}}{a^3} \tag{I.54}$$

We may note that according to Onsager the vector R coincides in direction with the vector of the total moment μ_{sum}. This leads to the situation that the reaction field causes only some increase in the moment of the dipole through induction and cannot change the rotation of the dipole, i.e., cannot orient the dipole.

Without giving all the calculations here, we may state that Onsager's theory connects the dielectric constant of a polar liquid with the dipole moment of its molecules by means of the following equation:

$$\frac{4\pi N}{3V} \cdot \frac{\mu_0^2}{3kT} = \frac{\left(\varepsilon_s - n^2\right)\left(2\varepsilon_s + n^2\right)}{\varepsilon_s \left(n^2 + 2\right)^2} \tag{I.55}$$

where n is the refractive index, V is the molar volume, and ε_s is static dielectric constant of the pure polar liquid. Thus, by measuring the dielectric constant, the refractive index, and the density of a pure polar liquid, it is possible to determine its dipole moment directly.

Onsager's theory is in good agreement with experimental results in the case of weakly polar liquids. But in the case of polar liquids with high dipole moments or liquids associated through hydrogen bonds it leads by calculation to values of the dielectric constant that are lower than the measured values. This is due to deviations from the theoretical assumptions upon which the derivation of equation (I.55) is based. The serious deficiencies of Onsager's theory are the following:

1. The model used by Onsager is an artificial and simplified model; in actual fact the distribution of the charges in the molecule of a polar liquid is considerably more complex.

2. The environment of a given molecule is considered as a continuous sphere with a constant macroscopic dielectric permittivity ε .

3. The interaction of a given molecule with its environment is considered only through the reaction field R which, being parallel to the direction of the dipole of the particular molecule, does not change its rotation. In other words, it does not take into account the fact that the orientation of the molecule under consideration is connected with the orientation of the neighboring molecules and, consequently, must depend on their arrangement.

The deficiencies of Onsager's theory were surmounted in Kirkwood's theory [18], in which in principle the interactions between the molecules were considered to be of arbitrary nature.

Kirkwood considers not the individual spherical molecule, like Onsager, but a whole spherical region of the dielectric containing a large number of molecules (the radius r of this sphere is sufficiently large in comparison with the dimensions of the molecules). In this case the surrounding sphere (external sphere of radius r) may be regarded with complete justification as continuous with a macroscopic dielectric constant ε_s.

The final expression to which Kirkwood's theory leads can be given in the form

$$\frac{(\varepsilon_s - 1)(2\varepsilon_s + 2)}{9\varepsilon_s} = \frac{4}{3}\pi N \left[\alpha_d + \frac{\mu_0^2 g}{3kT} \right] \qquad (I.56)$$

$$g = 1 + z\,\overline{\cos}\,\gamma$$

where z is the number of closest neighbors of a given molecule, i.e., the mean coordination number of a molecule of the polar liquid, and $\overline{\cos}\,\gamma$ is the mean value of the cosine of the angle between the directions of the dipoles of two neighboring molecules.

Kirkwood's theory was based on stricter initial assumptions than Onsager's theory. However, equation (I.56) includes the parameter g which cannot be determined experimentally, and it is therefore difficult to calculate the dipole moment from Kirkwood's formula. To find the magnitude g requires an accurate knowledge of the structure of the liquid and the nature of the forces of intermolecular interaction.

Kirkwood has made an attempt, using equation (I.56), to calculate the dielectric constant of water, using the value of the dipole moment in the gas. Assuming free rotation about the O—H bonds, the angle between which is 100°, he found $\overline{\cos}\,\gamma = 0.41$. The magnitude z can be found from structural data. For water z = 4. At $\overline{\cos}\,\gamma = 0.41$ (which corresponds to $\gamma = 100°$) the calculated value of the dielectric constant is 67, and at $\overline{\cos}\,\gamma = 0.5$ ($\gamma = 90°$) it is 82. A comparison of the latter magnitude for the dielectric constant with that found experimentally ($\varepsilon_s = 81$) shows the quality of Kirkwood's theory. However, similar calculations from equation (I.56) of the dielectric constant of aliphatic alcohols, which are also associated through intermolecular hydrogen bonds, give values

TABLE 1. Value of g for Aliphatic Ketones

Ketones	Dielectric permittivity at 20°C	Dipole moment μ_0, D	μ_0^2	$g\mu_0^2$	g
Acetone	21.45	2.85	8.12	9.82	1.21
Methyl ethyl ketone	18.51	2.77	7.67	9.97	1.30
Methyl propyl ketone	15.45	2.72	7.40	9.69	1.29
Diethyl ketone	17.00	2.71	7.34	10.40	1.42
Methyl isobutyl ketone	13.11	2.72	7.40	9.20	1.24
Methyl amyl ketone	11.95	2.70	7.29	9.10	1.25
Dipropyl ketone	12.60	2.72	7.40	9.60	1.30
Methyl hexyl ketone	10.39	2.70	7.29	8.70	1.19

close to those found by calculation from Onsager's formula and differing considerably from the experimental results. Thus, for ethanol the experimental figure is ε_e = 25.8, the figure calculated by Onsager's method is ε_p = 9.85, and that calculated by Kirkwood's method is ε_p = 9.67. For normal propanol, the respective figures are 22.2, 7.28, and 8.1.

Frenkel' [19] assumes that Kirkwood's calculations cannot be considered satisfactory either because it is difficult to imagine how the molecule rotates if it is simultaneously rigidly bound to its neighbors. In spite of the fact that it is difficult to determine the dipole moments of the molecule of a polar liquid directly by Kirkwood's theory (because of the indeterminacy of the parameter g), its validity can be confirmed by an independent method. Assuming that $g = 1 + z \overline{\cos} \gamma$, following Kirkwood, it must be considered that the magnitude g is determined by the structure and mutual arrangement of the polar molecules and consequently, must be similar in similar polar liquids. The results of calculation for a number of aliphatic ketones, according to Cole [20], are given in Table 1.

The Onsager—Kirkwood—Fröhlich statistical theory, which connects the static dielectric constant of polar liquids with the dipole moment, leads to the following general formula [5]:

$$g\mu_g^2 = \frac{(\varepsilon_s - \varepsilon_\infty)(2\varepsilon_s + \varepsilon_\infty)}{3\varepsilon_s}\left(\frac{3}{\varepsilon_\infty + 2}\right)^2 \frac{3vkT}{4\pi N_1} = \mu_l^2 \qquad (I.57)$$

TABLE 2. Values of μ_g and μ_l for a Number of Liquids [8]

Liquid	t°, C	ε_s	ε_∞	$d_4^{t^\circ}$	μ_g, D	μ_l, D	$\sqrt{g}$
Acetone	25	20.50	2.11	0.785	2.85	2.91	1.02
Nitrobenzene	25	35.25	2.43	1.198	4.21	4.20	1.00
Nitromethane	30	35.87	1.90	1.124	3.50	3.50	1.00
Pyridine	25	12.9	2.27	0.978	2.22	2.22	1.00
Chlorobenzene*	25	5.62	2.24	1.101	1.70	1,53	0.90
Chloroform	20	4.81	2.34	1.489	1.02	1.12	1.10
Ethyl ether	20	4.33	1.97	0.713	1.16	1.32	1.14

*In hexane, benzene, and carbon disulfide μ_g for chlorobenzene is 1.52-1.54 D, and $\sqrt{g}$ = 1.0.

where μ_g is the dipole moment in an infinitely rarefied gas and N_1 is the number of molecules in volume v. The parameter g has the same meaning as in Kirkwood's equation. Formula (I.57) is also approximate, since it is based on the assumption of a spherical form of the molecules of the polar dielectric and the continuity of a medium with dielectric constant ε_∞ up to the molecule under consideration. Consequently, deviations of the parameter g from unity may be caused not only by the ordering and orientation of the molecules (in the absence of the orientation effect of short-range forces, $\overline{\cos}\ \gamma = 0$, g = 1), but also by the imperfection of the theory itself [5]. Taking the latter into account, the parameter g may serve as a measure of the deviation of the effective dipole moment μ_l of the molecules in the liquid phase from the value of their moment μ_g in the gas phase:

$$\frac{\mu_l}{\mu_g} = \sqrt{g} \qquad (I.58)$$

Table 2 gives the results of calculations of μ_l from formula (I.57). It was drawn up with values of μ_g for a series of polar liquids.

For the first four liquids given in Table 2, there is good agreement of the values of μ_g and μ_l and the slight deviations of $\sqrt{g}$ from unity may be caused either by an inaccuracy in the measurements of the dielectric constant or, as mentioned above, by the imperfection of the theory. The establishment of a dependence of the value on the temperature makes it possible to draw a definite conclusion on the nature of the orientation of the molecules of a polar liquid [5]. For example, it has been established that for pyridine and nitrobenzene between 10 and 40°C $\sqrt{g}$ = 1. This shows that within the range of temperatures studied the arrangement of the dipoles of the molecules is random. The situation is different in the case of acetone. At temperatures above 10°C, $\sqrt{g}$ = 1, while below this temperature an increasing deviation of $\sqrt{g}$ from unity (in the direction of a decrease) is found, and therefore $\overline{\cos}\ \gamma$ < 0. This decrease in the parameter g at low temperatures is connected with the orientation of the dipoles of the acetone molecules in the opposite direction to one another.

Difficulties in the calculation of a dipole moment from equation (I.57) are connected primarily with finding the magnitude of ε_∞. It can be found from the formula

$$P_d = \frac{\varepsilon_\infty - 1}{\varepsilon_\infty + 2} V \tag{I.59}$$

where P_d is the deformation polarization, which is equal to the sum of the electronic and atomic polarizations, and V is the molar volume. The magnitude of ε_∞ can be determined by measuring the absorption or reflection of light in the far infrared region of the spectrum [21] and also from the results of measurements of dielectric constants and dielectric losses for polar liquids in the microwave range.

In those cases where experimental and calculated figures for the atomic polarization are lacking, P_d and R_D may be taken as equal and $\varepsilon_\infty \approx n_D^2$ [11].

Further investigations of the connection between the static dielectric constant of a polar liquid and the dipole moment of its molecule made by Fröhlich [4, 22] have led to the equation

$$\varepsilon_s - \varepsilon_\infty = \frac{3\varepsilon_s\,(2\varepsilon_s + \varepsilon_\infty)}{(2\varepsilon_s + 1)^2} \cdot \frac{4\pi N \vec{\mu}_i \vec{\mu}_i^*}{3kT} \tag{I.60}$$

$$\vec{\mu}_i \vec{\mu}_i^* = \vec{\mu}_i^2 (1 + z \cos \gamma)$$

where $\vec{\mu}_i$ is the dipole moment of the molecule of a polar liquid present in a medium with dielectric constant ε_s, and $\vec{\mu}_i^*$ is the dipole moment of a sphere containing a considerable number of molecules, having the dielectric constant ε_s, and surrounded by a di-

electric medium with the same dielectric constant ε_s. In contrast
to the dipole moment of a molecule in an infinitely rarefied gas
(vacuum dipole moment), the dipole moment $\vec{\mu}_i$ is not constant but
depends on ε_s and ε_∞. According to Fröhlich [4], this dependence
is expressed by the equation

$$\vec{\mu}_i = \frac{\varepsilon_\infty + 2}{3} \cdot \frac{2\varepsilon_s + 1}{2\varepsilon_s + \varepsilon_\infty} \vec{\mu}_g \qquad (I.61)$$

In Fröhlich's opinion [4], equation (I.60) is more general than
the equations (I.55)–(I.57) considered above, since its derivation is
not connected with the assumption of the spherical shape of the
molecules or with the continuity of a medium with dielectric constant ε_s right down to the molecule considered, which possesses a
point dipole at its center. Probably it must be considered that in
the absence of short-range forces the equality $\vec{\mu}_i = \vec{\mu}_i^*$ exists.

In spite of the fact that the derivation of equation (I.60) is not
based on the assumptions mentioned above, the results of calculations using it are not capable of unambiguous interpretation [5].
Thus, the value of $\overline{\mu_i \mu_i^*}$ depends on the temperature. For example,
for pyridine, nitrobenzene, and acetone $\overline{\mu_i \mu_i^*}$ decreases with a rise
in the temperature.

Table 3 gives the results of calculations of dipole moments
for a series of organic compounds from the equations of Debye
(I.49), Onsager (I.55), and Kirkwood (I.56)* in comparison with experimentally determined figures.

It follows from the figures in this table that Debye's equation
is unsuitable for calculating dipole moments of molecules of polar
liquids from the magnitude of their dielectric constants. The rea-

---

* For practical calculations it is convenient to use equations (I.49), (I.55), and (I.56)
in the following respective forms [8]:

$$\mu_0^2 = \frac{9kT}{4\pi N} \cdot \frac{M}{d} \left[\frac{\varepsilon - 1}{\varepsilon + 2} - \frac{\varepsilon_\infty - 1}{\varepsilon_\infty + 2} \right]$$

$$\mu_0^2 = \frac{9kT}{4\pi N} \cdot \frac{M}{d} \cdot \frac{(2\varepsilon + \varepsilon_\infty)(\varepsilon + 2)}{3\varepsilon(\varepsilon_\infty + 2)} \left(\frac{\varepsilon - 1}{\varepsilon + 2} - \frac{\varepsilon_\infty - 1}{\varepsilon_\infty + 2} \right)$$

$$g\mu^2 = \frac{9kT}{4\pi N} \cdot \frac{M}{d} \left[\frac{(\varepsilon - 1)(2\varepsilon + 1)}{9\varepsilon} - \frac{\varepsilon_\infty - 1}{\varepsilon_\infty + 2} \right]$$

TABLE 3. Dipole Moments of the Molecules of a
Number of Organic Compounds Calculated by the
Debye, Onsager, and Kirkwood Formulas

| Substance | μ_0, D (at 25°C) calculated according to | | | μ_{exp}^*, D |
	Debye	Onsager	Kirkwood	
Ethyl bromide	1.28	1.80	1.90	2.02 (g)
n-Propyl bromide	1.36	1.83	2.02	2.15 (g)
n-Butyl bromide	1.41	1.80	1.92	2.15 (g)
Isobutyl bromide	1.44	1.86	1.92	1.97 (l)
n-Amyl bromide	1.46	1.82	1.99	2.19 (g)
n-Hexyl bromide	1.51	1.88	1.99	1.97 (l)
n-Octyl bromide	1.56	1.81	1.88	1.96 (l)
n-Octyl chloride	1.58	1.84	2.14	—
n-Octyl iodide	1.46	1.65	1.80	1.89 (l)
Chlorobenzene	1.22	1.45	1.54	1.72 (g)
Bromobenzene	1.17	1.37	1.52	1.77 (g)
α-Chloronaphthalene	1.20	1.35	1.33	1.50 (l)
α-Bromonaphthalene	1.13	1.25	1.29	1.48 (l)

*g — gas, l — solution.

son for such considerable deviations in the magnitudes of the dipole moments calculated by equation (I.49) from experimental figures is, as has been shown above, the assumption in Debye's polarization theory that the internal field in a polar liquid corresponds to the Lorentz field.

The dipole moments calculated from Onsager's and Kirkwood's equations give results agreeing with the experimental figures, while Kirkwood's equation must be preferred.

Subsequently, in a number of papers attempts were made to improve the theory of the polarization of polar liquid dielectrics. The work in this direction was devoted mainly to extending Onsager's theory to an ellipsoidal model of the molecule of a polar substance [23-31]. In other papers, a number of empirical coefficients were introduced into Onsager's equation in order to improve it [32, 33, 47]. However, the modified equations obtained have no advantages whatever over the original Onsager equation.

The complexity of the equations given, the indeterminacy of the constants appearing in these equations, and the frequent discrepencies between the experimental and theoretical determinations hinder their use for calculating the dipole moments of molecules of polar liquids. From this point of view, Onsager's equation is the easiest to handle and the one most frequently used in practice.

A formula connecting the dielectric constant of a pure polar liquid with the dipole moment of its molecules has been proposed by Syrkin [34]:

$$\frac{\dfrac{M}{d}\cdot\dfrac{\varepsilon-1}{\varepsilon+2}-R}{1-\left(\dfrac{\varepsilon-1}{\varepsilon+2}\right)^2}=\frac{4}{3}\pi N\frac{\mu_0^2}{3kT} \tag{I.62}$$

In the majority of cases, the dipole moments calculated by means of formula (I.62) are in good agreement with the values obtained for dilute solutions [35]. Some theoretical justification of formula (I.62) has been given by Osipov and Shelomov [36].

An analogous formula for finding dipole moments of pure polar liquids has been proposed by Osipov [37]:

$$\left[\frac{(\varepsilon-1)(\varepsilon+2)}{8\varepsilon}-\frac{(n^2-1)(n^2+2)}{8n^2}\right]\frac{M}{d}=\frac{4}{3}\pi N\frac{\mu_0^2}{3kT} \tag{I.63}$$

The use of formula (I.63) to calculate the dipole moment of a molecule of a pure polar liquid from the magnitude of its dielectric constant has been illustrated with a large number of organic compounds [37], the error not exceeding +0.1 D. Thus, the dipole moments (in D) calculated from formula (I.63) for methyl iodide (1.41), chlorobenzene (1.55), pyridine (2.28), nitromethane (3.53), ethyl acetate (1.77), and phenylhydrazine (1.73) are extremely close to the moments of these compounds found for dilute benzene solutions: 1.41, 1.56, 2.25, 3.50, 1.81, and 1.70, respectively.

Formulas (I.62) and (I.63) can be used to determine the dipole moment of a polar substance in a polar solvent. In this case, (I.63) assumes the following form:

$$\frac{4}{3}\pi N\frac{\mu_1^2 x_1+\mu_2^2 x_2}{3kT}=P_{soln}^{or}=$$

$$=\left[\frac{M_1 x_1+M_2 x_2}{d_{1,2}}\right]\left[\frac{(\varepsilon_{1,2}-1)(\varepsilon_{1,2}+2)}{8\varepsilon_{1,2}}-\frac{(n_{1,2}^2-1)(n_{1,2}^2+2)}{8n_{1,2}^2}\right] \tag{I.64}$$

where $\varepsilon_{1,2}$, $n_{1,2}$, and $d_{1,2}$ are, respectively, the dielectric constant, the refractive index, and density of the solution, M_1 and M_2 are the molecular weights of the solvent and the solute, x_1 and x_2 are their mole fractions, and P_{soln}^{or}, the orientation polarization of the solution, is the sum of the orientation polarizations of the components. Knowing the magnitude P_{soln}^{or} it is possible to calculate the orientation polarization of a polar solute in a polar solvent from the expression

$$P_2^{or} = \frac{P_{soln}^{or} - P_1^{or}}{x_2} + P_1^{or} \qquad (I.65)$$

where P_1^{or} is the orientation polarization of the polar solvent calculated from formula (I.63). Formula (I.64) is applicable only to those cases in which the components of the system do not react chemically with one another.

Dipole moments (in D) have been calculated by means of formula (I.65) for methyl ethyl ketone and nitrobenzene in methyl benzoate (2.76 and 3.96, respectively) and for pyridine and quinoline in dimethylaniline (2.24 and 2.13, respectively) [38]. These magnitudes practically coincide with the moments obtained for benzene solutions of these compounds: 2.76, 3.96, 2.25, and 2.15, respectively.

The possibility of determining the dipole moments of a polar substance in a polar solvent extends the range of investigations in this direction, since very many polar substances are known which are extremely sparingly soluble in such nonpolar solvents as benzene, hexane, and the like.

4. Dielectric Properties of a Substance in a Variable Electric Field

Until now, we have discussed the behavior of a dielectric in a static electric field the frequency of oscillation of which is zero. The properties of a dielectric in such a field are described by the value of its static dielectric constant ε_s.

In the present section, we consider briefly the behavior of a dielectric substance in a variable electric field. It is obvious that the problem of creating a quantitative theory of the polarization of a dielectric in an electric field varying periodically with the time is more complex than in the case of a static field, since in the latter case it is not necessary to study the kinetic properties of the molecules.

We have already mentioned above that the polarization of a dielectric in an electric field possesses some inertia. When an external electric field is applied, the field of the molecular polarization of the dielectric reaches its static value not instantaneously but after a definite time. If the electric field is suddenly removed, the fall in polarization caused by the thermal motion of the molecules also takes place gradually. The thermal motion, causing a rearrangement of the orientations of the molecules, gradually returns the dielectric to its initial state, which corresponds to uniform distribution in the absence of an external electric field.

In the establishment of such a uniform distribution, the polarization of the dielectric will gradually decrease to zero. Its rate of fall with time can be described by some function a(t). If it is assumed that the dependence of the function a(t) on the time is subject to an exponential law, then we obtain [4, 5]:

$$a(t) = A_0 e^{-\frac{t}{\tau}} \tag{I.66}$$

Here A_0 and τ are constants which do not depend on the time but are functions of the temperature and the composition of the dielectric.

The constant τ is called the r e l a x a t i o n   t i m e. It characterizes the gradual change in the state of the dielectric under the polarizing influence of an electric field. More strictly, the relaxation time must be regarded as the time during which the polarization of the dielectric after the removal of the external field decreases by a factor of 1/e relative to its original value.

If, when the external field is removed, the rearrangement of the orientation of the molecules taking place under the action of the thermal motion is considered as its rotational changeover from one equilibrium position to another, in this process the molecule must overcome some potential energy barrier H_δ. The connection between H_δ and the relaxation time can be expressed by the following relation [4]:

$$\tau = \frac{\pi}{2\omega_a} A e^{\frac{H_\delta}{kT}} = \frac{\pi}{\omega_0} e^{\frac{H_\delta}{kT}} \tag{I.67}$$

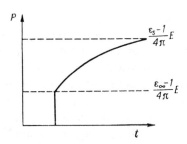

Fig. 4. The polarization P as a function of the time t on the instantaneous application of a constant field E.

where $\pi/2\omega_a$ is the mean time of reorientation of the excited molecule, A is a factor which changes slowly with the temperature, and $\omega_0/2\pi = \omega$ /πA is the frequency of the oscillations of the molecule about each equilibrium position.

In passing from a static field to the rapidly changing field of a light wave (with a vibrational period of the order of 10^{-14}-10^{-16} sec in the visible region), the molecular polarization $(\varepsilon_s - 1)$ E/4π falls, reaching the value $(\varepsilon_\infty - 1)$ E/4π (Fig. 4). The cause of the fall in the molecular polarization is that in rapidly changing fields the dipoles of the molecules of the dielectric, which possess a moment of inertia, are incapable of following the rapid changes in the direction of the field, in consequence of which the orientation polarization disappears completely, and the dielectric constant decreases from ε_s to ε_∞.

Let us pass to a short consideration of the dielectric properties of a substance placed in a variable electric field at intermediate frequencies.

Let us first consider an ideal dielectric condenser, assuming that when a voltage V is applied to its plates polarization sets in instantaneously. If this ideal condenser is placed in a variable electric field $E = E_0 e^{i\omega t}$, the dielectric displacement current (capacity current) will have a phase displacement of 90° with respect to the voltage.

In this case the capacity current will be

$$i = \omega V \varepsilon c_0 \tag{I.68}$$

where ω is the angular frequency, equal to 2π multiplied by the frequency in hertz, and c_0 is the capacity of the condenser in vacuum. In this ideal condenser, when the angle of phase displacement is 90°, no loss of electrical energy takes place.

The situation is different in the case of real dielectrics. Because of the phase lag in the orientation of the dipoles of the molecules of the dielectric with respect to the field and the existence of a finite period of dielectric relaxation, part of the electrical energy

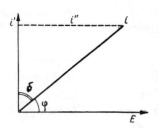

Fig. 5. Vector diagram of the active current i" and the re-active current i' for a conden-ser with losses.

is absorbed and dispersed in the form of heat. The current pass-ing through such a condenser now consists of two parts: the dielec-tric displacement current or ca-pacity current, which is $i' = \omega V \varepsilon' c_0$ and which is displaced in phase by 90° with respect to the applied variable voltage V, and the loss current or active current $i" = \omega V \varepsilon" c_0$, which is in the same phase as the voltage. The total current passing through the condenser is

$$\vec{i} = \vec{i'} + \vec{i"}$$ (I.69)

However, in a real dielectric, because of the absorption and dispersion of energy, the capacity current will lead the voltage by a phase angle of less than 90° (Fig. 5). The angle δ, the comple-mentary angle of the phase shift between the current strength and the voltage, is usually called the angle of dielectric loss.

We may note that the dispersion of energy in the dielectric is due to the active component of the total current, which coincides in phase with the applied voltage. In view of this, in the evaluation of the dielectric properties of a substance one must start from the ratio of the active current to the capacity current. In any case, a substance may be regarded as a dielectric if its active component does not greatly exceed its capacitive component.

The ratio of the loss current (active constituent) to the capa-city current is expressed by the tangent of the angle of dielectric loss and is an important characteristic of a dielectric

$$\tan \delta = \frac{i"}{i'} = \frac{\varepsilon"}{\varepsilon'}$$ (I.70)

Experimentally it is not the angle of dielectric loss that is usually determined but its tangent.

In relation (I.70), ε' is the measured dielectric constant of the substance filling the interelectrode space of the condenser and $\varepsilon"$ is the so-called dielectric loss factor.

Thus, if in a dielectric there is polarization constantly be-coming established, then in variable electric fields energy losses occur in it. In this case, the dielectric constant is the complex magnitude:

$$\varepsilon^* = \varepsilon' - i\varepsilon'' \tag{I.71}$$

where ε' and ε'' are called the real and imaginary components of the dielectric constant.

By measuring the capacity and resistance of a condenser containing a dielectric it is possible to determine ε' and ε'' from the following equations:

$$\varepsilon' = \frac{C}{C_0} \qquad \varepsilon'' = 1.8 \cdot 10^{12} \frac{\chi}{f}$$

where χ is the measured specific electrical conductivity for a variable current of frequency f.

The latter equation is equivalent to the relation $\varepsilon'' = 1/\omega R C_0$, where R is the resistance.

The dielectric constant ε' and the dielectric loss factor ε'' are functions of the frequency of the field. If the frequency of the variable field tends to zero ($\omega \to 0$), the behavior of the dielectric substance in it is similar to its behavior in a static field. Consequently, on the assumption that there are no losses in a constant field, we may write

$$\varepsilon''_{(\omega)} \to 0 \qquad \text{and} \qquad \varepsilon'_{(\omega)} \to \varepsilon_s \qquad \text{as} \qquad \omega \to 0$$

If the active component of the complex dielectric constant is not zero, $\varepsilon'_{(\omega)}$ will be somewhat less than the value of ε_s appearing in equations (I.49) and (I.55). On the other hand, at very high frequencies when $\omega \to \infty$, ε' tends to the magnitude of the optical dielectric constant ε_∞, i.e., $\varepsilon'_{(\omega)} \to \varepsilon_\infty$ at $\omega \to \infty$. These frequencies correspond to the infrared region of the spectrum [4].

The first equation for the frequency dependence of the complex dielectric constant was derived by Debye [1]:

$$\varepsilon^* = \varepsilon_\infty + \frac{\varepsilon_s - \varepsilon_\infty}{1 + i\omega\tau} \tag{I.72}$$

The simultaneous solution of equations (I.71) and (I.72) gives the following values for the real component ε' and the imaginary component ε'':

$$\varepsilon' = \varepsilon_\infty + \frac{\varepsilon_s - \varepsilon_\infty}{1 + \omega^2\tau^2} \tag{I.73}$$

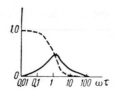

Fig. 6. Debye functions: Dashed line, $(\varepsilon' - \varepsilon_\infty)/(\varepsilon_s - \varepsilon_\infty)$; Solid line, $\varepsilon''/(\varepsilon_s - \varepsilon_\infty)$.

$$\varepsilon'' = \frac{(\varepsilon_s - \varepsilon_\infty)\,\omega\tau}{1 + \omega^2\tau^2} \tag{I.74}$$

Then, according to equation (I.70), for the tangent of the angle of dielectric loss we obtain

$$\tan\delta = \frac{\varepsilon''}{\varepsilon'} = \frac{(\varepsilon_s - \varepsilon_\infty)\,\omega\tau}{\varepsilon_s + \varepsilon_\infty\omega^2\tau^2} \tag{I.75}$$

Equations (I.73)–(I.75) describe the behavior of a dielectric in a variable electric field. Their derivation is based on the assumption that the process of establishing equilibrium is subject to the exponential law (I.66). In the majority of cases, this is valid when $\varepsilon_s - \varepsilon_\infty \ll 1$ [4]. This condition is satisfied by dilute solutions. The real and imaginary components of the complex dielectric constant depend not only on the frequency but also on the temperature. However, their temperature dependence does not appear in clear form, but through the temperature-dependent relaxation time and the difference $\varepsilon_s - \varepsilon_\infty$.

If the Debye equations given above are valid, the functions $(\varepsilon' - \varepsilon_\infty)/(\varepsilon_s - \varepsilon_\infty)$ and $\varepsilon''/(\varepsilon_s - \varepsilon_\infty)$ arising from the relations may be illustrated graphically by the curves shown in Fig. 6.

At constant temperature, at the maximum the angular frequency is defined by the condition $\partial\varepsilon''/\partial\omega = 0$, $\omega_{max} = \omega$. At the maximum value of the loss ε'', $\omega_{max}\tau = 1$. Substituting this value in equations (I.73)–(I.75), we obtain

$$\varepsilon'_{max} = \frac{\varepsilon_s + \varepsilon_\infty}{2} \tag{I.76}$$

$$\varepsilon''_{max} = \frac{\varepsilon_s - \varepsilon_\infty}{2} \tag{I.77}$$

and for the tangent for the angle of dielectric loss:

$$\tan\delta = \frac{\varepsilon_s - \varepsilon_\infty}{\varepsilon_s + \varepsilon_\infty} \tag{I.78}$$

These equations satisfactorily describe (qualitatively) the changes in the dielectric constant and dielectric loss, regardless of the mechanism of the processes taking place in dielectrics. For a number of simple substances, the quantitative changes also agree with the experimental results [39].

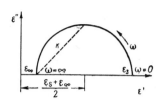

Fig. 7. Circular diagram for the complex dielectric constant ε^*.

It follows from equation (I.74) that the dielectric loss factor approximates to zero for both small and large values of $\omega\tau$. This conclusion corresponds to the ideas discussed above on the lagging of orientation polarization behind the field.

Debye's formulas can be used for the analysis and graphical representation of the experimental results. The simultaneous solution of equations (I.73) and (I.74) leads to the expression

$$\left(\varepsilon' - \frac{\varepsilon_s - \varepsilon_\infty}{2}\right)^2 + (\varepsilon'')^2 = \left(\frac{\varepsilon_s - \varepsilon_\infty}{2}\right)^2 \tag{I.79}$$

which is the equation of a circle with radius $R = (\varepsilon_s - \varepsilon_\infty)/2$. If, now, the values of ε'' are plotted along the axis of ordinates and the values of ε' as a function of the frequency ω along the axis of abscissas, the points corresponding to Debye's equations will lie on a semicircle with a center at $(\varepsilon_s + \varepsilon_\infty)/2$ (Fig. 7).

There are data in the literature confirming the applicability of the Debye formulas. Thus, the semicircles constructed from the experimental data for n-propanol [40], nitrobenzene [5], a number of monosubstituted benzenes C_6H_5X [41] (X = F, Cl, Br, C, N), and other compounds agree well with the Debye theory taking only one relaxation time into account.

According to equation (I.77), at the point of the maximum ε_{max}, the orientation polarization decreases to half its original value and the frequency ω_{max} corresponding to this point is equal to $1/\tau$. Consequently, when Debye's formulas are satisfied (in the case of a single relaxation time) the relaxation time and $(\varepsilon_s - \varepsilon_\infty)/2$ can be determined from graphical data (cf. Fig. 7).

However, the relaxation times of the molecules of polar dielectrics are not all the same, as a rule, and therefore there is a distribution of relaxation times. The range of distribution of relaxation times is particularly wide for highly polymeric compounds [42, 43].

Various functions and methods have been proposed for calculating the distribution of relaxation times [42, 44]. The equation

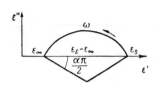

Fig. 8. Cole–Cole diagram.

derived by Cole and Cole [45] has been the most successful and most widely used for representing the experimental results:

$$\varepsilon' = \varepsilon_\infty + \frac{\varepsilon_s - \varepsilon_\infty}{1 + (i\omega\tau_0)^{1-\alpha}} \qquad (I.80)$$

where α is an empirical factor representing the distribution parameter and assuming values of 0 to 1 and τ_0 is the most probable relaxation time. It corresponds to the frequency ω_{max} at which $\varepsilon'' = \varepsilon''_{max}$ and relative to which the other relaxation times are distributed symmetrically. At $\alpha = 0$, equation (I.80) becomes the Debye equation.

In the Cole–Cole method, the semicircle corresponding to one relaxation time is generally replaced by the arc of a circle the center of which lies below the actual axis (Fig. 8). A diameter of this circle cuts the axis of abscissas at the point corresponding to the value of ε_∞ and forms with the axis of abscissas the angle $\pi\alpha/2$. By plotting through the experimental points obtained for various frequencies an arc the center of which lies below the ε' axis, it is possible to find the values of α and τ_0.

The first theory of relaxation polarization was proposed by Debye [1]. He derived an equation providing the possibility of determining the relaxation time for a spherical molecule of radius *a* rotating in a liquid with the macroscopic viscosity η:

$$\tau = \frac{4\pi\eta a^3}{kT} \qquad (I.81)$$

Formula (I.81) is approximate, since the macroscopic viscosity is used in place of the microscopic viscosity and the molecule is regarded as a sphere moving in a continuous viscous medium. Perrin [46] has generalized Debye's theory for ellipsoidal molecules.

Literature Cited

1. P. Debye, Polare Molekeln, Leipzig (1929) [English translation, The Chemical Catalogue Co., New York (1929)].
2. P. Debye and H. Sack, Die Theorie der elektrischen Eigenschaften von Molekeln [Russian translation], ONTI, Moscow (1936).
3. W. Brown, Dielectrics [Russian translation], IL, Moscow (1961).

4. H. Fröhlich, The Theory of Dielectrics, Clarendon Press, Oxford (1949).
5. M. I. Shakhparonov, Methods of Studying the Thermal Motion of Molecules and the Structure of Liquids, Izd. MGU (1963).
6. G. I. Skanavi, The Physics of Dielectrics (Region of Weak Fields), Gostekhizdat (1949).
7. A. R. von Hippel, Dielectric Materials and Applications, MIT and Wiley, New York (1954).
8. C. P. Smyth, Dielectric Behavior and Structure, McGraw-Hill, New York (1955).
9. C. J. Böttcher, Theory of Electric Polarisation, Elsevier, Amsterdam (1952).
10. J. H. Van Vleck, The Theory of Electric and Magnetic Susceptibilities, Oxford University Press, Oxford (1932).
11. R. J. LeFevre, Dipole Moments, Methuen, London (1948).
12. J. W. Smith, Electric Dipole Moments, Butterworths, London (1955).
13. A. Weissberger (Editor), Physical Methods in Organic Chemistry (Vol. I of Technique of Organic Chemistry), Interscience, New York, 2nd Edn. (1949).
14. B. V. Ioffe, Refractometric Methods in Chemistry, Goskhimizdat, Moscow (1960).
15. P. Langevin, J. Phys., 4:678 (1905); Ann. Chim. Phys., 5:70 (1905).
16. A. I. Ansel'm, Zh. Éksperim. i Teor. Fiz., 12:274 (1942); 13(11):432 (1943).
17. L. Onsager, J. Am. Chem. Soc., 58:1486 (1936).
18. J. Kirkwood, J. Chem. Phys., 7:911 (1939).
19. Ya. I. Frenkel', The Kinetic Theory of Liquids, Izd. Akad. Nauk SSSR (1945).
20. R. H. Cole, J. Chem. Phys., 9(3):251 (1941).
21. F. H. Cartwright and J. Errera, Proc. Roy. Soc., A154:138 (1936).
22. H. Fröhlich, Trans. Faraday Soc., 44:238 (1938).
23. C. V. Raman and K. S. Krishnan, Proc. Roy. Soc., 117A:589 (1939).
24. J. N. Wilson, Chem. Rev., 25:377 (1939).
25. G. Jaffe, J. Chem. Phys., 6:385 (1938).
26. M. Jasumi and H. Komooka, Bull. Chem. Soc. Japan, 29:407 (1956).
27. H. Fröhlich and R. Sack, Proc. Roy. Soc., 182A:388 (1944).
28. G. B. Brown, Nature, 150:661 (1942).
29. S. K. K. Jatkar, B. K. Y. Jyenar, and N. V. Sathe, J. Indian Inst. Sci., 28A:1 (1946).
30. S. K. K. Jatkar and B. K. Y. Jyenar, J. Indian Inst. Sci., 30A:27 (1948).
31. R. Wecver and R. W. Rany, Inorg. Chem., 5:703 (1966).
32. J. H. Van Vleck, J. Chem. Phys., 5:566 (1937).
33. P. C. Henriquez, Rec. Trav. Chim., 54:574 (1935).
34. Ya. K. Syrkin, Dokl. Akad. Nauk SSSR, 35:45 (1942).
35. A. N. Shidlovskaya and Ya. K. Syrkin, Zh. Fiz. Khim., 22:913 (1948).
36. O. A. Osipov and I. K. Shelomov, 30:608 (1956).
37. O. A. Osipov, Zh. Fiz. Khim., 31:1542 (1957).
38. O. A. Osipov and M. A. Panina, Zh. Fiz. Khim., 32:2287 (1958).
39. The Physics and Chemistry of the Solid State of the Organic Compounds [Russian translation], Izd. Mir, Moscow (1967), Chap. 12.
40. D. W. Davidson and R. H. Cole, J. Chem. Phys., 19:1484 (1951).
41. I. P. Poley, J. Appl. Sci. Res., 4:337 (1955).
42. W. Kauzmann, Rev. Mod. Phys., 14:12 (1942).
43. R. M. Fuoss and J. C. Kirkwood, J. Am. Chem. Soc., 61:385 (1941).

44. D. H. Branin and C. P. Smyth, J. Am. Chem. Soc., 74:1121 (1952).

45. K. S. Cole and R. H. Cole, J. Chem. Phys., 9:341 (1941).

46. F. Perrin, J. Phys. Rad., 5(7):497 (1937).

47. W. R. Gilherson and K. K. Srivasfuves, J. Phys. Chem., 64:1485 (1960).

48. O. A. Osipov and V. I. Minkin, Handbook of Dipole Moments, Izd. "Vysshaya shkola" (1965).

Chapter II

Nature of the Dipole Moment and Methods for Its Determination

1. Nature of the Dipole Moment

As has been mentioned above, the electric dipole moment is defined by the relation

$$\vec{\mu} = q\vec{l}$$

where $\vec{l}$ is the radius vector directed from the center of gravity of the negative electric charge to the center of gravity of the positive electric charge, the absolute magnitude of each charge being $q*$ [1, 2].

If $\rho(\vec{r})$ is the charge density of the molecule, the magnitude

$$\vec{\mu} = \int \rho(\vec{r})\,\vec{r}\,dr \qquad \text{(II.1)}$$

is called the dipole moment of the molecule. We may write the charge density of the molecule $\rho(\vec{r})$ in the form of the sum of $\rho_{el}(\vec{r})$, the density of the electronic charge, and $\rho_n(\vec{r})$, the density of the charge connected with the nuclei:

---

* It must be noted that in the chemical literature and in the present book the opposite direction of the dipole moment vector is taken. In order of magnitude the dipole moment is equal to the electronic charge ($4.8 \cdot 10^{-10}$ esu) multiplied by a distance of 1 Å (10^{-8} cm), which amounts to $4.8 \cdot 10^{-18}$ esu. It is customary to measure the dipole moments in Debye units: one Debye is equal to $1 \cdot 10^{-18}$ esu.

$$\rho\,(\vec{r}) = \rho_n(\vec{r}) + \rho_{el}\,(\vec{r}) \qquad (II.2)$$

Since the total charge of an un-ionized molecule is zero,

$$\int \rho_n(\vec{r})\,dv = -\int \rho_{el}\,(\vec{r})\,dv = en \qquad (II.3)$$

where n is the number of electrons in the molecule and e is the elementary charge. The dipole moment may be written as

$$\vec{\mu} = q\,(\vec{R}^+ - \vec{R}^-) \equiv q\vec{l} \qquad (II.4)$$

Here the following symbol q = ne has been introduced:

$$\vec{R}^+ = \frac{1}{q}\int \vec{r}\rho_n(\vec{r})\,dv = \frac{1}{q}\sum e_a\vec{r}_a \qquad (II.5)$$

where e_a is the charge of a nucleus located at a point with the radius vector $\vec{r}_a$.

$$\vec{R}^- = \frac{1}{q}\int \vec{r}\rho_{el}\,(\vec{r})\,dv; \qquad \vec{l} = \vec{R}^+ - \vec{R}^-$$

The magnitudes $\vec{R}^+$ and $\vec{R}^-$ are the radii vectors of the "centers" of the positive and negative charges and $\vec{l}$ is the radius vector directed from the center of the negative charges to the center of the positive charges.

The dipole moment of a complex molecule can be represented approximately in the form of the vectorial sum of the moments belonging to the individual bonds (see Chapter III) and then the problem of the appearance of polarity in a molecule reduces to elucidating the electric asymmetry of the individual bonds.

The simplest case of the chemical bond isolated from the influence of neighboring bonds is that of a diatomic molecule, which can conveniently be used as an example for a consideration of the factors causing the appearance of an electric dipole moment [3, 4].

According to the LCAO MO method, the atoms A and B in the diatomic molecule AB act as centers of the bonding molecular orbital:

$$\Psi = N\,(\varphi_A + \lambda\varphi_B) \qquad (II.6)$$

Here φ_A and φ_B are the valence atomic orbitals belonging to atoms A and B, N is a normalizing factor, and λ is a numerical factor

determining the contribution of each atomic orbital and characterizing the ionicity of the bond. For a covalent bond formed by atoms
of the same electronegativity, $\lambda = 1$. The greater the difference in
the electronegativities of A and B, the more does λ differ from
unity.

The density of the negative charge at any point is defined as

$$\rho = |\psi|^2 = \frac{(\varphi_A + \lambda\varphi_B)^2}{1 + \lambda^2 + 2\lambda s} \tag{II.7}$$

where $s = \int\limits_{-\infty}^{+\infty} \varphi_A\varphi_B \, dv$ is the overlap integral. The numerical val-

ues of these integrals have been calculated for various types of
atomic orbitals and are given in tables [5]. By selecting the origin
of coordinates at the central point O at which the center of gravity
of the positive charge is located, it is possible to define the solution
of the center of gravity of the negative charge C (Fig. 9) by its coordinates

$$\bar{x} = \int x\rho \, dv = \frac{\int x(\varphi_A + \lambda\varphi_B)^2 \, dv}{1 + \lambda^2 + 2\lambda s} = \frac{\int x\varphi_A^2 \, dv + 2\lambda \int x\varphi_A\varphi_B \, dv + \lambda^2 \int x\varphi_B^2 \, dv}{1 + \lambda^2 + 2\lambda s} \tag{II.8}$$

The functions $\int x\varphi_A^2 dv$ and $\int x\varphi_B^2 dv$ are the coordinates of the
centers of the electron density for the atomic orbitals φ_A and φ_B.
If these orbitals are not hybridized, their centers are obviously located on the nuclei A and B, i.e., they have the coordinates $-R/2$
and $R/2$.

Let us assume that the numerical value of $\int x\varphi_A\varphi_B dv$ is negligibly small. In actual fact, the product $\varphi_A\varphi_B$ differs appreciably
from zero only in the region between A and B where x is small.
Then expression (II.8) reduces to

$$\bar{x} = \frac{(\lambda^2 - 1) R}{1 + \lambda^2 + 2\lambda s} \tag{II.9}$$

and since $\mu = 2e\bar{x}$, the dipole moment is

$$\mu = \frac{(\lambda^2 - 1) eR}{1 + \lambda^2 + 2\lambda s} \tag{II.10}$$

Expression (II.10) characterizes that part of the polarity of
the bond that is a consequence of the difference in the electronegativities of the atoms forming it. In actual fact, the more λ differs
from unity the greater is the absolute magnitude of the dipole moment μ. At $\lambda = 1$, $\mu = 0$. However, the difference in the electro-

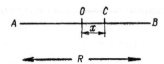

Fig. 9. Determination of the center of gravity of the negative charge in a diatomic molecule AB.

negativities of the atoms is not the only cause of the appearance of a dipole moment. Even in the case of a purely covalent bond ($\lambda = 1$) a nonzero dipole moment μ may arise because of the member

$$\frac{2\lambda \int x\varphi_A\varphi_B \, dv}{1 + \lambda^2 + 2\lambda s}$$ in formula (II.8).

This component of the total dipole is called the homopolar dipole or overlap moment. As Mulliken and Coulson [6, 7] have shown, the magnitude of the homopolar dipole is the greater the greater the difference in the effective dimensions of the atomic orbitals forming the bond. Coulson and Rogers [8] have carried out numerical calculations of homopolar dipoles for many types of bonds and diatomic molecules. For example, for the H−Cl bond formed by the 1s orbital of hydrogen and the 3p orbital of chlorine, the homopolar dipole is 1.4 D with the positive center on the chlorine atom. Figure 10 illustrates schematically the origin of a homopolar dipole. Because of the different shapes of the s and p orbitals (the latter is considerably elongated), the center of gravity of the overlap charge ($\varphi_A\varphi_B$) is displaced in the direction of the hydrogen by $\bar{x}_h$. The magnitude $\mu_h = 2e\bar{x}_h$ and is the homopolar dipole.

Where the bonding and nonbonding orbitals are hybridized, which is the case in almost all actual molecules, the centers of the electron densities of the hybrid orbitals are no longer located on the nuclei and $\int x\varphi_A^2 dv$ and $\int x\varphi_B^2 dv$, their coordinates, do not coincide with $-R/2$ and $R/2$. Thus, if φ_A is a hybridized orbital of the type $\varphi_A = N(s + \chi p)$, where χ is a factor characterizing the weight of the p orbital, the coordinates of the center of electron density of the φ_B orbital will be

$$\frac{R}{2} - \frac{1.78\chi a_0}{1 + \chi^2}$$

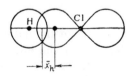

Fig. 10. Scheme of the origin of a homopolar dipole.

where a_0 is the radius of the Bohr orbit.

The additional dipole moment due to the term $1.78_\chi \, a_0/(1 + \chi^2)$ is called the hybridization atomic dipole. The value of atomic dipoles is fairly large. Thus, the

atomic dipole calculated for the one electron forming the sp^3 orbital of carbon is 1.89 D; for sp^2: 2.07 D; and for sp: 2.20 D.

The atomic dipoles arising through the hybridization of nonbonding electronic orbitals are particularly large. Thus, the dipole
moment of ammonia (1.46 D) is due mainly to the moment of the
nonbonding pair of electrons of the nitrogen atom which is, as
shown by calculation, 3.67 D. The values of the atomic dipoles depend very strongly on the magnitude of the hybridization parameter,
which cannot be determined experimentally. Consequently, the calculation of atomic dipoles is extremely indeterminate and is sensitive to the assumption introduced [9].

Thus, the appearance of a dipole moment is due to the following main factors:

1. The displacement of the center of gravity of the charge of
the bonding electrons towards the more highly electronegative atom
of the bond.

2. The appearance of a homopolar dipole as a consequence
of the difference in the dimensions of the atomic orbitals forming
the bonding molecular orbital.

3. The asymmetry of the atomic orbitals forming the bonding
molecular orbital which arises because of their hybridization.

4. The asymmetry of a nonbonding pair of electrons due to
hybridization.

In conclusion, it must be emphasized that although the scheme
of the separation of the dipole moment into its constituents given
above is extremely helpful, it has one fundamental disadvantage.
As Ruedenberg has observed [10], the relative contributions of the
various factors responsible for the total dipole moment depend to
an extreme degree on the selection of the center of coordinates.
Ruedenberg derived an equation for dipole resolution which is
invariant with respect to the coordinate system. The components
of the dipole moment that he has obtained are comparable in the
physical sense with those discussed above.

2. Methods of Determining Dipole Moments

The practical determination of a dipole moment is based on
the existence of an orientation effect (except for the molecular
beam method) of polar molecules in an applied electric field. The

calculated magnitudes of dipole moments may be regarded as reliable only where intermolecular interaction is excluded and the molecule can freely orient itself in the applied field. From this point of view, the most reliable data on dipole moments can be obtained by carrying out studies in the gas phase at very low pressures when the distance between the molecules is so large that electrostatic interaction between them is almost absent.

The existing experimental methods of determining dipole moments of molecules can be divided arbitrarily into two groups. The first group of methods is based on measuring the dielectric constant. This includes, in the first place, the methods proposed by Debye for determining dipole moments in vapors and in dilute solutions of polar substances in nonpolar solvents. This group may also include methods connected with the measurement of the dielectric constant of individual polar liquids with subsequent calculation of the dipole moments by Onsager's formulas and other equations considered in the first chapter.

The second group includes methods based on microwave spectroscopy and molecular beams: the Stark method, the resonance microwave method, the molecular beam method, etc.

Of all the methods of determining dipole moments, the most widely used are those based on measurements of the dielectric constant in vapors and dilute solutions of polar substances in nonpolar solvents. The majority of experimental values of dipole moments has been obtained by these methods [11–14] which are based on Debye's statistical theory [15–16].

3. Determination of the Dipole Moment in the Vapor Phase

The determination of the dipole moments of molecules in the vapor state is performed in accordance with the Debye formula

$$P = \frac{4}{3} \pi N \left(\alpha_d + \frac{\mu^2}{3kT} \right)$$

which can be given in the form

$$P = a + \frac{b}{T} \quad \text{or} \quad PT = aT + b \tag{II.11}$$

where $a = {}^4/_3 \pi N \alpha_d = P_e + P_a$ and $b = {}^4/_9 \pi N \frac{\mu^2}{k} = P_{or} T$.

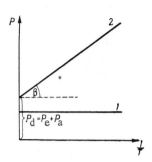

Fig. 11. The polarization P as a function of $1/T$: 1) Nonpolar substance; 2) polar substance.

It follows from formula (II.11) that the molecular polarization P is a linear function of $1/T$.

To find the permanent dipole moment in the vapor state, the dielectric constant and the density of the substance under investigation are measured at several temperatures, the corresponding values of the molecular polarization P are found from formula (I.49), and these are plotted on a graph as a function of $1/T$. Then the tangent of the angle of slope of the straight line obtained gives the value of b and extrapolation of the line to $1/T = 0$ gives the intercept P_d on the axis of ordinates which corresponds to the magnitude $P_d = P_e + P_a$. The larger the dipole moment of the molecule, the larger the angle of slope. In the case of a nonpolar molecule (b = 0), the molecular polarization does not depend on the temperature and the straight line 1, as the function $P = f(1/T)$, is parallel to the axis of abscissas (Fig. 11).

From the angle of slope it is possible to calculate the dipole moment of a polar molecule from the following simple equation:

$$\mu = 0.01283 \sqrt{\tan \beta} \cdot 10^{-18} \text{ esu} \qquad (II.12)$$

The measurements of the dielectric constant and density must be carried out at the lowest possible pressures. In this case it is possible to assume that the mutual polarizing action of the molecules of a polar substance in the absence of an applied field are extremely small and, consequently, the local field must, to a first approximation, coincide with the applied external field.

However, at higher pressure the action of a field on a molecule under consideration may be considerably modified by the polarization and orientation of the surrounding molecules, and then the function $P = f(1/T)$ will deviate from linearity because of intermolecular interaction.

The Debye method considered is based on the assumption that the permanent dipole moment of a molecule does not depend

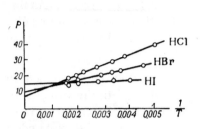

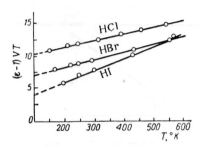

Fig. 12. Polarization P as a function of 1/T for gaseous HCl, HBr, and HI.

Fig. 13. The magnitude $(\varepsilon - 1) VT$ as a function of T for gaseous HCl, HBr, and HI.

on the temperature. Consequently, the method cannot be applied to substances in which the state of the molecules changes with the temperature.

The accuracy of the determination of the dipole moment by the method given is governed by the accuracy with which the angle of slope of the straight line $P = f(1/T)$ is found, and this requires the measurements of the dielectric constant and density in the vapor phase to be carried out over a wide temperature range, of the order of $100°$ and more, which is not always possible.

This method cannot be used to determine the dipole moment of individual polar liquids, since the distances between the molecules in a polar liquid are so small that they may be comparable with the dimensions of the molecules themselves and such adjacent molecules will exert an extremely considerable polarizing action on one another, which will lead to large deviations of the function $P = f(1/T)$ from linearity. In other words, the method is applicable only to those cases in which the molecules possess completely free orientation in an external field, i.e., have a random distribution outside the field.

In applying Debye's equation to gases, it is frequently simplified by assuming that $(\varepsilon + 2) = 3$. The possibility of this simplification is based on the fact that the dielectric constant of gases is extremely close to unity. Then equation (I.49) assumes the form

$$(\varepsilon - 1) V = 4\pi N \left(\alpha_d + \frac{\mu^2}{3kT}\right) \qquad (II.13)$$

where V is the molar volume of the gas.

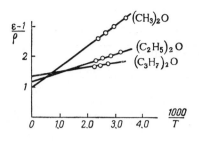

Fig. 14. The magnitude $(\varepsilon - 1)/\rho$ as a function of 1000 / T for gaseous methyl, ethyl, and propyl ethers at constant density ρ and moderate pressures [74].

An analysis of equation (II.13) shows that not only is the molecular polarization P a linear function of $1/T$ but the magnitude $(\varepsilon - 1)$ VT depends linearly on the temperature. In place of the molar volume it is possible to use any magnitude proportional to it such as the specific volume or the density.

One of the first attempts at the experimental verification of the Debye equation was Zahn's [17] study of the dipole moments of the hydrogen halides HCl, HBr, and HI in the gas phase. Figure 12 shows the molar polarizations of these compounds as a function of $1/T$, and Fig. 13 shows $(\varepsilon - 1)$ VT as a function of the temperature. In Fig. 13, V represents the ratio of the specific volume of a mass of the gas measured at the temperature of the experiment to the specific volume at 273°K. Both volumes are reduced to atmospheric pressure and the measurement of ε was also carried out at atmospheric pressure.

As follows from Figs. 12 and 13 [44], all the experimental points lie satisfactorily on straight lines. The extrapolation of these lines to intersect the axis of ordinates leads to the following values for the dipole moments: for HCl 1.034, for HBr 0.788, and for HI 0.382 D. These values agree well with the results of the determination of the dipole moments by the Stark method (1.04, 0.79, and 0.38 D, respectively).

Sänger et al. [18] have measured the dielectric constants and densities ρ of methyl, ethyl, and propyl ethers in the gas phase and found a linear relation between the functions $(\varepsilon - 1)/\rho$ and 1000/T (Fig. 14). The polarizabilities and constant dipole moments of all three ethers were determined from these curves and are given in Table 4.

In the method considered, the temperature measurements of the dielectric constant can be carried out by two methods. In the first, constant pressure is preserved which requires an accurate knowledge of the dependence of the density of the gas on the tem-

TABLE 4. Polarizabilities and Dipole
Moments of Ethers in the Gas Phase [18]

Compound	Polarizability $\alpha \cdot 10^{24}$, cm^8	Dipole moment, D
Methyl ether	6.1	1.29
Ethyl ether	11.0	1.11
Propyl ether	16.6	1.00

perature. In the second method, the density of the gas remains unchanged and the dielectric constant and the pressure of the gas are measured, the latter then being recalculated for an ideal gas or the polarization at zero pressure is found by extrapolation.

The dielectric constants of a large number of polar substances in the vapor form have been studied, and in all cases the molar polarization was found to be a linear function of $1/T$. Exceptions were formed by some compounds in which a change in the temperature led to a change in either the molecular state or in the value of the permanent dipole moment (for example, 1,2-dihaloethanes). Thus, an experimental test of the Debye equation has shown that it is strictly satisfied if the investigation of the permanent dipole moments is carried out in the vapor phase at low pressures.

We may note that measurements of the temperature dependence of the molecular polarization in the gas phase make it possible to determine the atomic polarization. As already mentioned, the intercept P_d (see Fig. 11) corresponds to the deformation polarization $P_d = P_e + P_a$. Knowing the value of the electronic polarization P_e it is possible to calculate the atomic polarization P_a as the difference $P_d - P_e$.

This method of determining polarization is accurate in the case of measurements in the gas phase, but leads to high values with measurements in liquids, which can be explained by a change in the solvent effect as a function of the temperature [11, 19].

The atomic polarization can be determined indirectly also. If we measure the dielectric constant of a given substance in the solid state at a temperature lower than the melting point and calculate the molecular polarization by formula (I.49), because of the absence of an orientation effect under these conditions the molecular polarization obtained will be $P_e + P_a = P_d$. Hence, knowing the electronic polarization P_e it is possible to calculate the atomic polarization P_a.

Both methods of determining the atomic polarization are associated with great practical difficulties. The second method may lead to considerable errors because of the complexity of the accurate determination of the specific gravity of a solid substance and also because of the fact that the small value of the atomic polarization to be determined is the difference between two comparatively large polarization figures ($P_d - P_e$) having a number of sources of error in their measurement.

The method of determining dipole moments in the gas phase from the temperature dependence has a more considerable theoretical basis and is free from systematic errors. However, it possesses some disadvantages of a methodical nature: the comparative complexity of the apparatus used, its laboriousness, and the impossibility, in many cases, of converting the substance from the condensed phase into the vapor phase without decomposition. In practice, measurements of the dielectric constant in the gas phase over a wide range of temperatures can be carried out only for comparatively simple and volatile compounds.

Another far less accurate but experimentally more manageable method for determining a dipole moment in the gas phase exists. It is used in those cases where the dielectric constant is measured at one temperature and several pressures. From the results obtained the values of the molecular polarization P_{mol} are calculated and these figures are extrapolated to zero pressure in order to eliminate the electrostatic interaction between polar molecules.

The essence of the method consists in the fact that the first member of the Debye equation

$$P = \frac{4}{3} \pi N \alpha_d + \frac{4}{9} \pi N \frac{\mu^2}{kT}$$

characterizing the deformation polarization ($P_e + P_a$) is identified with the molecular refraction extrapolated to infinitely long waves R_∞ (frequently R_D is chosen as approximately equivalent to R_∞). Essential for the theoretical basis of the method is the fact that at optical frequencies the dipoles of polar molecules are unable to become orientated and, consequently, the term $\frac{4}{9}\pi N\mu^2/kT$ in the Debye equation that depends on the temperature must disappear. Then

$$P_{mol} - R_D = P_{or}$$

Thus, knowing the values of the molecular polarization and the molecular refraction, it is possible to find the orientation polarization and, from the formula

$$P_{or} = \frac{4}{3}\pi N \frac{\mu^2}{3kT}$$

to calculate the constant dipole moment of the molecule:

$$\mu = 0.01283\sqrt{P_{or}\,T}\cdot 10^{-18}\ \text{esu} \tag{II.14}$$

In this method no account is taken of atomic polarization and the magnitude a, the deformation polarization $P_e + P_a$ in formula (II.11), is equated with P_e (R_D). Of course, this introduces a certain error in the determination of the dipole moment. The error may be considerable where the orientation polarization is low. Consequently, dipole moments of the order of 0.4-0.5 D determined by this method must be regarded as approximate. Large moments (2-5 D) can be determined with an accuracy of ± 0.1 D.

4.   Determination of the Dipole Moment
in Dilute Solutions (Debye's Second Method)

The method of determining the dipole moment of a molecule in the gas phase which is based on the temperature dependence of the molecular polarization cannot be applied to dilute solutions of a polar substance in nonpolar solvents since the solvent effect varies with the temperature and leads to a deviation of the function $P = f(1/T)$ from linearity.

According to Debye [15], the second method of finding the constant dipole moment of a molecule in the gaseous state considered above can be applied to dilute solutions with satisfactory approximation. In this connection, Debye starts from the assumption that the internal field of the solution is the Lorentz field. Consequently, in extremely dilute solutions the molecules of polar substances must behave just as in the vapor state and must become freely oriented in the applied field. It is clear that this assumption cannot be strictly based.

Assuming additivity of the properties of the components of the solution, we can write the relation

$$\frac{\varepsilon_{1,2}-1}{\varepsilon_{1,2}+2} = \frac{4}{3}\,\pi\,(N_1\alpha_1 + N_2\alpha_2) \tag{II.15}$$

where α_1 and α_2 are the polarizabilities of the molecules of the solvent and of the solute, respectively, N_1 is the number of molecules of the solvent in 1 cm^3, N_2 is the number of molecules of solute in 1 cm^3, and $\varepsilon_{1,2}$ is the static dielectric constant of the solution.

Introducing the mole fractions

$$x_1 = \frac{N_1}{N_1+N_2} \quad x_2 = \frac{N_2}{N_1+N_2}$$

and taking into account the fact that the molar polarizations of the solvent and the solute are equal, respectively, to

$$P_1 = \frac{4}{3}\,\pi N\alpha_1 \quad \text{and} \quad P_2 = \frac{4}{3}\,\pi N\alpha_2$$

the ratio (II.15) can be given in the form

$$\frac{\varepsilon_{1,2}-1}{\varepsilon_{1,2}+2} \cdot \frac{M_1 x_1 + M_2 x_2}{d_{1,2}} = P_1 x_1 + P_2 x_2 \tag{II.16}$$

where M_1 and M_2 are the molecular weights of the solvents and solute and $d_{1,2}$ is the density of the solution.

The left-hand side of equation (II.16) is the molar polarization of the solution $P_{1,2}$, and consequently

$$P_{1\cdot2} = P_1 x_1 + P_2 x_2 \tag{II.17}$$

Since the polarization of the solution $P_{1,2}$ and the polarization of the solvent P_1 are determined directly from the experimental values of the dielectric constant and the density, and the mole fractions x_1 and x_2 are given, it is easy to calculate the values of the molar polarization of the solute P_2 from formula (II.17):

$$P_2 = \frac{P_{1,2}-P_1}{x_2} + P_1 \tag{II.18}$$

In formula (II.18), the polarization P_1 is taken as equal to the polarization of the pure solvent, which is not strictly correct, since this does not take into account the interaction (mainly of the deformation type) between the molecules of the solute and of the solvent.

As a rule, P_2 increases with a decrease in the concentration of the solute, the rate of increase also having a tendency to rise with dilution.

Experimentally, the method of determining a dipole moment in dilute solutions is simpler and more convenient. In practice, the experiment reduces to measuring the dielectric constant and the density of 4-6 dilute solutions of the compound under investigation at a single temperature, for example over the range of concentrations of solute from 0.001 to 0.01 or from 0.02 to 0.10 mole fractions [20]. The selection of the range of concentrations is governed by considerations concerning the required accuracy of the measurements and also by the solubility of the substance under study in the given solvent. Then the values of the molar polarization of the solute for each concentration are found from formula (II.18) and the results obtained are used to plot a curve of P_2 as a function of the concentration. To exclude the effect of a residual interaction between the molecules of the solute, P_2 is extrapolated to zero concentration ($x_2 = 0$) and the value of the molar polarization at infinite dilution, $P_{2\infty}$, is found:

$$P_{2\infty} = \lim_{x_2 \to 0} P_2 = \lim_{x_2 \to 0} \left(\frac{P_{1,2} - P_1}{x_2} + P_1 \right) \tag{II.19}$$

Knowing the value of $P_{2\infty}$, it is easily possible to determine the orientation polarization of the solute

$$P_{2\infty}^{or} = P_{2\infty} - P_e - P_a$$

and from formula (II.14) the constant dipole moment is calculated as follows:

$$\mu = 0.01283 \sqrt{(P_{2\infty} - P_e - P_a)T} \cdot 10^{-18} \text{ esu} \tag{II.20}$$

Smyth considers that in the determination of $P_{2\infty}$ the error connected with the influence of the solvent may amount to 10%. As a rule, it leads to a decrease in the value of $P_{2\infty}$ [20]. In view of this, the neglect of P_a which is frequently practiced does not lead to a serious error in calculations of the dipole moment (if $\mu > 1$ D) since it is less than the error introduced through the solvent effect and, moreover, has the opposite sign.

The experimental results show that in the majority of cases the error connected with the solvent effect is less than 10% and moreover is almost the same in the determination of the dipole moments of dissolved compounds in one and the same solvent and even in related solvents. The dipole moments of aliphatic ketones determined in benzene solutions are between 2.7 and 2.8 D and the dipole moments of nitrobenzene in hexane, heptane, and decalin are, respectively, 4.05, 3.93, and 4.1 D [12, 21].

As mentioned above, the sum $P_e + P_a$ in formula (II.20) can be replaced by the molecular refraction R_D (for the sodium yellow line). Then

$$\mu = 0.01283 \sqrt{(P_{2\infty} - R_D)T} \cdot 10^{-18} \text{ esu} \tag{II.21}$$

When there are no conjugated bonds in the molecule of the substance under investigation (absence of exaltation of the refraction), the molecular refraction may, with an accuracy which is satisfactory for dipole moments, be calculated by the additive method from the atomic refractions [22, 23]. The value of R_D can be found directly from measurements of the refractive index and the density of the substance under investigation in the pure state, since the dependence of R_D on the concentration is so slight that it can easily be neglected. In the case of solid substances, however, the molecular refraction is found from measurements of the refractive index and of the density of solutions using a formula analogous to that for polarization:

$$R_{1,2} = \frac{n_{1,2}^2 - 1}{n_{1,2}^2 + 2} \cdot \frac{M_1 x_1 + M_2 x_2}{d_{1,2}}$$

$$R_2 = \frac{R_{1,2} - R_1}{x_2} + R_1$$

where $n_{1,2}$, $d_{1,2}$, and $R_{1,2}$ are, respectively, the refractive index, density, and molecular refraction of the solution, R_1 is the molecular refraction of the solvent, M_1 and M_2 are the molecular weights of the solvent and the solute, and x_1 and x_2 are their molar fractions.

The extrapolation of the polarization P_2 to infinite dilution is not accurate because of the curvilinear nature of the function $P_2 = f(x_2)$. In view of this, attempts have been made to decrease the errors by using extrapolation formulas.

Hedestrand [24] has proposed, instead of the polarization of the solute, to extrapolate directly to zero concentration ($x_2 \to 0$) the dielectric constant and the density of the solution which, in the majority of cases, are linear functions of the concentration in molar fractions of the solute:

$$\varepsilon = \varepsilon_1 (1 + \alpha x_2) \tag{II.22}$$

$$d = d_1 (1 + \beta x_2) \tag{II.23}$$

where ε_1 and d_1 are, respectively, the dielectric constant and the density of the solvent and α and β are coefficients characterizing the slopes of the straight lines describing the dependence of ε and d on x_2. The coefficients α and β can also be determined by the method of least squares.

From these relations, Hedestrand obtained the following equations for the polarization of the solute at infinite dilution:

$$P_{2\infty} = \frac{\varepsilon_1 - 1}{\varepsilon_1 + 2} \cdot \frac{M_2}{d_1} + \frac{M_1}{d_1} \frac{[3\alpha\varepsilon_1 - \beta (\varepsilon_1 - 1) (\varepsilon_1 + 2)]}{(\varepsilon_1 + 2)^2} \tag{II.24}$$

where M_1 and M_2 are, respectively, the molecular weights of the solvent and the solute.

Equation (II.24) cannot be used if the dependences of the dielectric constant and the density on the concentration of the solute are nonlinear, which is frequently the case in concentrated solutions.

The method of analytical extrapolation that has been considered has some advantages over the method in which $P_{2\infty}$ is found from the dependence of P_2 on the concentration, since, in the first place the extrapolation carried out is linear and, in the second place it enables the errors arising as a result of the contamination of the solvent impurities to be reduced. The dielectric constant of the solvent can be found by extrapolating the dependence of the dielectric constant of the solution on the concentration of solute to $x_2 = 0$. If there is a substantial difference, exceeding the experimental error, between the values of the dielectric constant of the solvent determined by extrapolation and that measured directly, it must be assumed that impurities are present in solution. In particular, during an experiment the solvent may absorb water vapor, which leads to an increase in its dielectric constant.

Hedestrand's formula is one of the most successful formulas for analytical extrapolation and has been widely used. Some modifications were later introduced into Hedestrand's method. For example, the use of the logarithmic differentiation of the Clausius —

Mosotti function has led to the following equation:

$$P_2 - P_1 = \frac{3P_1}{\varepsilon_1^2 + \varepsilon_1 - 2} \left(\frac{\Delta\varepsilon}{x_2}\right)_{x_2 \to 0} - \frac{P_1}{d_1}\left(\frac{\Delta d}{x_2}\right)_{x_2 \to 0} + \frac{P_1}{M_1}(M_2 - M_1) \qquad (\text{II.25})$$

Here and below, the index 1 relates to the solvent, the index 2 to the solute, and the values without indices to the solution.

In this case the magnitude $\Delta\varepsilon/x_2$ is extrapolated to zero concentration of the solute, its dependence on x_2 being more linear than that of the dielectric constant itself. The orientation polarization P_2^{or} of the substance under investigation is found by deducting the molecular refraction from P_2 determined by one of the methods mentioned above.

An analogous formula was obtained by Robles [25]:

$$P_2^{\text{or}} = A\left[\left(\frac{\Delta\varepsilon}{x_2}\right)_{x_2=0} - \left(\frac{\Delta n_D^2}{x_2}\right)_{x_2=0} + BM_2\right]. \qquad (\text{II.26})$$

where A and B are constants depending on the nature of the solvent and the temperature.

Le Fevre and Vine [26], starting from the linear dependences of the dielectric permeability and the density on the weight fraction of the solute ω_2,

$$\varepsilon = \varepsilon_1(1 + \alpha\omega_2) \qquad (\text{II.27})$$

$$d = d_1(1 + \beta\omega_2) \qquad (\text{II.28})$$

have introduced an equation for the polarization of the solute at infinite dilution

$$P_{2\infty} = \frac{\varepsilon_1 - 1}{\varepsilon_1 + 2} \cdot \frac{1}{d_1}(1 - \beta) + \frac{3\alpha\varepsilon_1}{d_1(\varepsilon_1 + 2)} = P(1 - \beta) + c\alpha \qquad (\text{II.29})$$

where P and c are constants characterizing the solvent.

Le Fevre considers that the dielectric constant and the density of a solution may not be linear functions of the weight fraction of the solute if the molecular weights of the solvent and the solute are unequal [27]. In view of this, in the Le Fevre−Vine method it is best to extrapolate the values of the coefficients α and β to zero concentration.

Halverstadt and Kumler [28] have used a linear dependence of the dielectric constant and the specific volume v of the solution on the weight fraction of the solute

$$\varepsilon = \varepsilon_1 + \alpha\omega_2 \qquad (\text{II.30})$$

$$v = v_1 + \beta \omega_2 \qquad \text{(II.31)}$$

and from these relations have obtained an equation for the polarization at infinite dilution

$$P_{2\infty} = \frac{3 a v_1 M_2}{(\varepsilon_1 + 2)^2} + M_2 \, (v_1 + \beta) \, \frac{\varepsilon_1 - 1}{\varepsilon_1 + 2} \qquad \text{(II.32)}$$

Equation (II.32) has been widely adopted, since the use of the concentration expressed in weight fractions of the solute considerably simplifies the calculations. We may note that when d = d_1, equation (II.32) is identical with the Le Fevre –Vine equation.

Fujita [29], using instead of the function v = $f(\omega_2)$ the dependence of the density on ω_2, obtained the following equation for $P_{2\infty}$:

$$P_{2\infty} = \frac{\varepsilon_1 - 1}{\varepsilon_1 + 2} \cdot \frac{M_2}{d_1} \left[1 + \frac{3 a}{(\varepsilon_1 - 1)(\varepsilon_1 + 2)} - \frac{\beta}{d_1} \right] \qquad \text{(II.33)}$$

where α and β are the coefficients of the equations

$$\varepsilon = \varepsilon_1 + a\omega_2 \qquad \text{(II.34)}$$

$$d = d_1 + \beta\omega_2 \qquad \text{(II.35)}$$

As can be seen, equation (II.34) is a modification of the Hedestrand equation.

On the basis of the assumption that a specific polarization of the components of the mixture at any concentration can be found by the intercept method, Gross [30] has developed a method for curvilinear analytical extrapolation. However, Gross's method has not found application because of the necessity of using an extremely wide range of concentrations.

As can be seen from what has been said, the object of the extrapolation equations considered above was to find the total polarization of the solute at infinite dilution $P_{2\infty}$, from which the orientation polarization $P_{2\infty}^{or}$ was then calculated. Subsequently, a number of equations have been proposed which relate the experimentally determined magnitudes directly to the orientation polarization.

Guggenheim [31], starting from the assumption made in the derivation of the Debye equation that the interaction between the molecules of the solute in dilute solutions is negligibly small, considered the polarization and refraction of a solution as additive magnitudes. For this case, the Debye formula can be brought into the following form:

$$P_{2\infty}^{or} + P_a = \left(\frac{\varepsilon-1}{\varepsilon+2} - \frac{\varepsilon_1-1}{\varepsilon_1+2}\right) - \left(\frac{n^2-1}{n^2+2} - \frac{n_1^2-1}{n_1^2+2}\right) = \frac{4\pi N}{3}\left(\frac{\mu^2}{3kT} + \alpha_a + \alpha_a'\right)C_2$$

$$\text{(II.36)}$$

where C_2 is the concentration of the solute in moles per unit volume. According to Guggenheim, α_a is the atomic polarizability of the solvent multiplied by the ratio of the molecular volumes of the solute and the solvent v_2/v_1, and α_a' is some fictitious value of the atomic polarizability which corresponds to α_a for the case where the atomic polarizabilities of the solvent and the solute are proportional to their molar volumes. The magnitude α_a' is defined by formula (II.36):

$$\frac{4}{3}\pi N\alpha_a' = V_2\left(\frac{\varepsilon_1-1}{\varepsilon_1+2} - \frac{n_1^2-1}{n_1^2+2}\right) \qquad \text{(II.37)}$$

In the limiting case $(C_2 \to 0)$, and when there is no interaction between the molecules of the solute, equation (II.36) can be given in the form

$$P_{2\infty}^{or} + P_a = \frac{3M_2V_1}{(\varepsilon_1+2)^2}\lim_{\omega_2\to0}\frac{d\varepsilon}{d\omega_2} - \frac{3M_2V_1}{(n_1^2+2)^2}\lim_{\omega_2\to0}\frac{dn^2}{d\omega_2}$$

$$= \frac{4}{3}\pi N\left(\frac{\mu^2}{3kT} + \alpha_a + \alpha_a'\right) \qquad \text{(II.38)}$$

Since there are no experimental data for atomic polarization, it is determined by calculation for the three approximate relations: 1) $\alpha_a = \alpha_a'$; 2) $\alpha_a = \alpha_a'/2$; and 3) $\alpha_a = 3\alpha_a'/2$.

Where the atomic polarization is neglected, equation (II.38) assumes the simpler form

$$P_{2\infty}^{or} = \frac{3M_2V_1}{(\varepsilon_1+2)^2}\left(\frac{d\varepsilon}{d\omega_2}\right)_{\omega_2\to0} - \frac{3M_2V_1}{(n_1^2+2)^2}\left(\frac{dn^2}{d\omega_2}\right)_{\omega_2\to0} = \frac{4}{3}\pi N\frac{\mu^2}{3kT} \qquad \text{(II.39)}$$

But if it is considered that the difference between the dielectric constant and the refractive index of the solvent is very small, equation (II.39) can be simplified still further:

$$P_{2\infty}^{or} = \frac{3M_2V_1}{(\varepsilon_1+2)^2}\left[\left(\frac{d\varepsilon}{d\omega_2}\right)_{\omega_2\to0} - \left(\frac{dn^2}{d\omega_2}\right)_{\omega_2\to0}\right] = \frac{4}{9}\pi N\frac{\mu^2}{kT} \qquad \text{(II.40)}$$

Using equation (II.40) it is easy to calculate the dipole moment of the solute [32, 33]:

$$\mu^2 = \frac{10^{36}}{N}\cdot\frac{9kT}{4\pi}\cdot\frac{3}{(\varepsilon_1+2)^2}\cdot\frac{M_2}{d_1}\left(\frac{\Delta}{\omega_2}\right)_{\omega_2\to0} \qquad \text{(II.41)}$$

where $\Delta = (\varepsilon - \varepsilon_1) - (n^2 - n_1^2)$.

In Guggenheim's method, the determination of the density of the solution is necessary only to calculate the concentration. Palit [34] has introduced some refinements into the Guggenheim equation. He has derived a complete equation for the orientation polarization:

$$P_{2\infty}^{or} = \frac{3\left(\varepsilon_1 - n_1^2\right)}{d_1\left(\varepsilon_1 + 2\right)\left(n_1^2 + 2\right)}\left(1 - \frac{\beta}{d_1}\right) + \frac{3\alpha}{d_1\left(\varepsilon_1 + 2\right)^2} - \frac{6n_1 V}{d_1\left(n_1^2 + 2\right)^2} \quad (II.42)$$

Smith [35], using the linear dependence of the dielectric constant and the refractive index on the concentration expressed in weight fractions, has proposed an equation analogous to the Guggenheim equation:

$$P_{2\infty}^{or} = \frac{3M_2 V_1 \lim_{\omega_2 \to 0} \dfrac{d\left(\varepsilon - n^2\right)}{d\omega_2}}{\left(\varepsilon_1 + 2\right)^2} \quad (II.43)$$

In the Smith method, measurement of the density of the solution is eliminated completely. Thus, in the Guggenheim−Palit−Smith method the measurement of the density is replaced by the measurement of the refractive index of the solutions. This has some advantage over other methods of analytical extrapolation since the measurement of the density in the determination of dipole moments is a most laborious operation.

Subsequently, attempts were made to decrease to the limit the number of parameters to be determined experimentally for calculating the dipole moments of solutes. Thus, Higasi [36] proposed an equation in which all parameters except the dielectric constant were completely absent:

$$\mu = B\sqrt{\frac{\Delta\varepsilon}{C_2}} \quad (II.44)$$

where B is a constant depending on the nature of the solvent and the temperature. For benzene at 25°C, B = 0.90 ± 0.1, and for hexane it is 1.15 ± 0.1. Somewhat later, Bal and Srivastava [37] gave a more accurate value of B for benzene (0.828). The Higasi equation can be substantiated theoretically [38].

Srivastava and Charandas [39] have proposed an analogous equation in which the concentration of solute is expressed in weight fractions:

$$\mu = A\sqrt{M_2\alpha} \quad (II.45)$$

where $A = 0.012812\sqrt{\dfrac{3T}{d_1\,(\varepsilon_1+2)^2}}$ (T is the absolute temperature).

Srivastava and Charandas [39] have changed the value of the coefficient A somewhat by introducing an empirical correction. At ordinary temperatures, for benzene A = 0.090.

None of the extrapolation formulas considered above are exact and, as follows from the literature [41, 44, 45], they are approximately equivalent for calculating dipole moments in very dilute solutions of a polar substance in a nonpolar solvent. However, as already mentioned, the Hedestrand and Halverstadt−Kumler formulas have been the most widely used. Possible approximations and assumptions in the calculation of dipole moments have been discussed repeatedly in the literature [19, 40, 45]. Sutton et al. have given a detailed discussion of the required degree of accuracy in the measurement and calculation of the various parameters used for finding the dipole moment [40]. To determine a moment with an accuracy of not less than ± 0.01 D the coefficients α and β and also the dielectric constant, the refractive index, and the density, must be measured with an accuracy of about 0.0001. Particular attention must be devoted to the accuracy of the determination of the coefficient α, which introduces a large error. The most accurate results for the coefficients α and β are obtained by using the method of least squares for their calculation. It is possible to use simple formulas to evaluate the errors of each of the parameters mentioned above when calculated by the method of least squares. For example, for the coefficient α [46-48]:

$$\Delta\alpha = \pm\sqrt{\frac{n}{n-2}\cdot\frac{\sum(\varepsilon'-\varepsilon)^2}{n\sum\omega_2^2-\left(\sum\omega_2\right)^2}}\qquad\text{(II.46)}$$

where n is the number of measurements, and ε' represents the "best" values of the dielectric constant.

Dipole moments of not less than 1.5 D can be determined with an accuracy of about 1%. For example, Böttcher [44] calculated the dipole moment of phenol from benzene solutions by the Debye, Hedestrand, and Guggenheim methods and obtained, respectively, 1.59, 1.60, and 1.59 D. As can be seen, the difference is less than 1%.

This agreement of the results does not mean that the dipole moment of phenol has actually been determined with an accuracy of 1%, since in all the methods the calculations are based on the Debye formula [41]. However, the results obtained indicate the compatibility of the values calculated by the use of formulas (II.21), (II.24), and (II.36).

As the dipole moment decreases, the effect of errors in the measurements and calculations increases considerably because the orientation polarization is a function of the square of the dipole moment. Consequently, a certain circumspection must be used in the interpretation of dipole moments of the order of 0.4-0.5 D obtained by the dilute solution method [41].

In view of the approximate nature of the Debye formula, attempts have been made to use Onsager's molecular model and its modifications to calculate the dipole moment from results obtained on dilute solutions [41]. For practical use, Onsager's equation (I.60) for solutions of a polar substance in a nonpolar solvent can be brought into various forms, for example,

$$\mu^2 = \varepsilon_1 - n_1^2 \left(2 + \frac{n_1^2}{\varepsilon} \right) \frac{9kT}{4\pi N} \cdot \frac{M}{d \left(n_1^2 + 2 \right)^2} \tag{II.47}$$

or

$$\mu^2 = \left[\left(\varepsilon - n_1^2 \right) - V_2 \frac{\left(n_2^2 - n_1^2 \right) 3n_1^2}{\left(n_2^2 - n_1^2 \right)} \right] \frac{9kT}{4\pi N} \tag{II.48}$$

Scholte [49] has extended Onsager's theory to an ellipsoidal model of a polar molecule and has derived an equation for calculating the dipole moment in nonpolar solvents. This equation in a somewhat modified form can be written as

$$P = \frac{(\varepsilon_1 - 1)(2\varepsilon_1 + 1)}{12\pi\varepsilon_1 N} \cdot \frac{M}{d} \left[1 + \frac{2\varepsilon_1^2 + 1}{\varepsilon_1 (\varepsilon_1 - 1)(2\varepsilon_1 + 1)} \left(\frac{\partial \varepsilon}{\partial \omega_2} \right)_{\omega_2 \to 0} - \frac{1}{d_1} \left(\frac{\partial d}{\partial \omega_2} \right)_{\omega_2 \to 0} \right] \tag{II.49}$$

$$\mu^2 = 3kT \left(1 - f\alpha_2 \right) \left(P - \frac{\alpha_2}{1 - f\alpha_2} \right) \tag{II.50}$$

where f, the internal field factor, expressed in terms of the semi-axis of the molecular ellipsoid, is also a function of the dielectric constant.

On the basis of experimental results for solutions, Scholte calculated the dipole moments of chloroform, chlorobenzene, and nitrobenzene by means of the Debye and Onsager formulas (II.49)

TABLE 5. Dipole Moments of Chloroform, Chlorobenzene, and
Nitrobenzene Calculated by Various Methods

Compound	Debye formula	Onsager formula	Scholte formula	In the gas phase
Chloroform	1.1	1.1	1.0	1.0
Chlorobenzene	1.6	1.5	1.7	1.7
Nitrobenzene	3.9	3.8	4.3	4.2

and compared the results obtained with the values found for the
gas phase (Table 5).

In all three methods, extrapolation to zero concentration of
the solute was used. However, the formulas used for calculating
the dipole moment were different since they are based on different
theoretical assumptions. While the Debye formula is based on the
Lorentz theory of the local field and takes no account of the presence of the reaction field, the Onsager formula is based on a spherical molecular model and the Scholte formula takes into account
the anisotropy of the polarizability of the polar molecule.

The Scholte method gives values of the calculated moments
best agreeing with the figures obtained in the gas phase (Table 5).
It can clearly be seen from these examples that the use of the Onsager spherical model cannot introduce any improvements whatever
into the determination of dipole moments unless the anisotropy of
the polarizability of the polar molecule and its geometry are taken
into account [31].

Ross and Sack [50], just like Scholte, have used an ellipsoidal
model of the polar molecule. The equation that they derived for
solutions can be given in the form

$$\varepsilon - \varepsilon_1 = \frac{1}{3} v_0 N (\varepsilon_0 - \varepsilon_1) \sum \frac{1}{\varepsilon_1 + D_s (\varepsilon_0 - \varepsilon_1)} + \frac{4\pi}{3} N \frac{\mu^2}{3kT} \qquad (\text{II}.51)$$

where ε_0 is the assumed microscopic dielectric constant of the ellipsoid, v_0 is the volume of the ellipsoid, which is equal to $\frac{4}{3}\pi abc$
(where a, b, c are the semiaxes of the ellipsoid), and D_s is the so-
called depolarizing factor. For a spherical molecule, $D_s = \frac{1}{3}$.
Somewhat later, we shall show that equation (II.51) can be used to
calculate the solvent effect.

In conclusion, we may note that the methods of calculating dipole moments based on the direct extrapolation to zero concentration of the solute of the dielectric constant and the density give a smaller scatter of the values obtained than methods in which the polarization is subjected to extrapolation. For example, the dipole moment of nitrobenzene (nitrobenzene is frequently used as the test substance) calculated from benzene solutions in four series of measurements had the following values when the dielectric constant was extrapolated: 4.00, 3.98, 3.93, and 4.01 D; and the values 3.97, 4.08, 3.93, and 3.94 D when the polarization was extrapolated. The mean values of the moment of nitrobenzene calculated by the two methods are in good agreement (3.99 and 3.98 D [51]). Böttcher [44, 52], using the Onsager formula, has calculated the dipole moments of molecules of a number of organic and inorganic compounds from results on the dielectric constants and refractive indices of the liquids at infinite wavelengths ($\lambda \to \infty$) and has obtained results differing little from the moments in the gas phase. Böttcher considers that the Onsager formula can be used successfully for calculating dipole moments of pure liquids with the exception of those cases in which the molecules are associated through hydrogen bonds or possess hindered rotation [44].

5. Solvent Effect

Müller [53-55] was the first to show that the dipole moments calculated from the experimental results for dilute solutions by means of the Debye formula differ from the moments for substances in the gas phase. This phenomenon has acquired the name of the solvent effect. The main cause of the existence of the solvent effect is that the Debye formula, the derivation of which is based on the theory of the local Lorentz field, cannot take into account the electrostatic interaction arising between molecules of the solute and the molecules of the solvent surrounding it. As a rule, the dipole moments determined by the dilute solution method are somewhat lower than those found in a gas or vapor.

In a number of investigations, attempts to calculate a theoretical correction for the solvent effect have been made; reviews of these investigations are given in the literature [45, 56, 57]. A number of semiempirical and empirical formulas have also been proposed of the general type [19]:

$$\mu_s = \mu_0 \left[1 + cf(\varepsilon_1)\right] \qquad (\text{II}.52)$$

where μ_0 and μ_s are the dipole moments in the gas phase and in solution, respectively, ε_1 is the dielectric permeability of the solvent, and c is a constant.

Müller [55] studied the influence of 25 different individual solvents and nine mixed solvents on the dipole moment of chlorobenzene and came to the conclusion that the discrepency between the dipole moment in the vapor and that measured in solution increased with an increase in the dielectric constant of the solvent. He showed that in the range of dielectric constants from 1.8 to 2.9 the ratio μ_s/μ_0 can be expressed by the following relation:

$$\mu_s = \mu_0 \left[1 + c(\varepsilon_1 - 1)^2\right] \qquad (\text{II}.53)$$

where c is an empirical constant equal to 0.038.

Apart from equation (II.53), the following expressions have been proposed for the function $f(\varepsilon_1)$: $(\varepsilon_1 - 1)$ [58]; $(\varepsilon_1 - 1)/(\varepsilon_1 + 2)$ [59]; $3/(\varepsilon_1 + 2)$ [60]; $1/\varepsilon_1$ [61].

Both Müller's equation and the expressions for the function $f(\varepsilon_1)$ considered are approximate and therefore give more or less satisfactory results for certain classes of compounds and prove unsuitable for other classes.

Guggenheim [62], using the equation

$$\frac{\left(\varepsilon - \varepsilon_1 - n^2 + n_1^2\right)C}{\varepsilon_1} = \frac{4\pi\mu^2}{3kT} \qquad (\text{II}.54)$$

where C is the number of moles of solute in 1 cm^3 of solution, proposed an approximate equation for taking the solvent effect into account in the form

$$\frac{\mu_s \text{ [Guggenheim]}}{\mu_0 \text{ [Debye]}} = \frac{\varepsilon_1 + 2}{3\sqrt{\varepsilon_1}} \qquad (\text{II}.55)$$

The dipole moment in solution was calculated from formula (II.55). For ordinary nonpolar solvents with a dielectric constant of the order of 1.8-2.3, the ratio μ_s/μ_0 is 0.942-0.947.

Raman and Krishna's equation taking into account the anisotropy of the polarizability of the molecules was used by Govinda Rau [63] to calculate the solvent effect. The equation that he obtained can be expressed in the form:

$$P_{2\,(gas)} = P_{2\,\infty} - N_0\psi_1 \frac{3a\varepsilon_1}{(\varepsilon_1+2)^2} - \frac{\varepsilon_1-1}{\varepsilon_1+2} N_0\psi_2 - \frac{\varepsilon_1-1}{\varepsilon_1+2}\cdot\frac{N_0\theta_2}{3kT} \qquad (\text{II}.56)$$

where ψ_1 and ψ_2 are functions taking into account the anisotropy of the polarizability of the molecules and θ_2 is the anisotropy of the constant dipole moment. The magnitude a is defined by the following formula:

$$\varepsilon = \varepsilon_1\,(1+ax_2)$$

Govinda Rau used his equation to calculate the dipole moment of nitrobenzene in various solvents and obtained satisfactory results. However, Raman and Krishna's theory has been criticized in the papers of Müller [64] and of Jenkins and Bauer [65]. In the opinion of these authors, equation (II.56) gives only the order of magnitude of the solvent effect in the majority of cases and is therefore unsuitable for accurate calculations.

Higasi [66] and Frank [67], independently of one another, came to the same conclusion that the solvent effect depends on the shape of the molecule and the charge distribution in it.

According to Higasi, the apparent moment in solution μ_s is

$$\mu_s = \mu_0 + \mu_i \qquad (\text{II}.57)$$

where μ_0 is the dipole moment measured in the gas phase, and μ_i is the effective moment arising from the interaction of the moments of the solvent and the solute.

Higasi's final equation for taking the solvent effect into account may be written in the form

$$\frac{\mu_s}{\mu_0} = 1 + 3A\,\frac{\varepsilon_1-1}{\varepsilon_1+2} \qquad (\text{II}.58)$$

where A is a constant depending on the shape and dimensions of the solute molecule. Equation (II.58) is similar to Sugden's empirical formula for which the function $f(\varepsilon_1)$ is expressed by the ratio $(\varepsilon_1-1)/(\varepsilon_1+2)$.

Starting from the same assumptions as Higasi, Frank derived an analogous equation but with slightly different values of the function $f(\varepsilon_1)$ and the coeeficient A [67]:

$$\frac{\mu_s}{\mu_0} = 1 + A'\,\frac{\varepsilon_1-1}{\varepsilon_1} \qquad (\text{II}.59)$$

Here, A' is some numerical magnitude equal to the sum $A_1 + A_2$, where A_1 depends only on the shape of the molecule of solute and

TABLE 6. Dipole Moments of Some Organic Compounds

Compound	Dipole moment, D		μ_s/μ_0		
	gas	benzene	experimental results	from equation (II.57)	from equation (II.60)
CH_3Cl	1.86	1.64	0.88	0.95	0.97
CH_3Br	1.78	1.78	1.00	0.94	0.97
CH_3I	1.59	1.58	1.00	0.99	1.00
$CHCl_3$	1.08	1.21	1.12	1.08	1.05
C_2H_5Br	2.02	1.89	0.94	0.94	0.96
C_2H_5I	1.90	1.78	0.94	0.89	0.93
C_6H_5F	1.57	1.45	0.92	0.96	0.97
C_6H_5Cl	1.70	1.56	0.92	0.91	0.94
C_6H_5Br	1.73	1.51	0.88	0.92	0.95
$C_6H_5NO_2$	4.23	3.98	0.94	0.89	0.96
C_6H_5CN	4.39	3.94	0.90	0.92	0.95
HCN	2.93	2.59	0.88	0.85	0.91
$n-C_4H_9CN$	4.09	3.57	0.87	0.87	0.92
CH_3NO_2	3.54	3.13	0.88	0.95	0.97
CH_3COCH_3	2.85	2.76	0.96	0.96	0.98
$C_6H_5COCH_3$	3.00	2.95	0.98	0.98	0.99
$(C_2H_5)_2O$	1.16	1.22	1.06	1.05	1.04
CH_3OH	1.69	1.62	0.96	1.01	1.01
C_2H_5OH	1.69	1.66	0.98	1.02	1.01
$n-C_3H_7OH$	1.66	1.66	1.00	1.04	1.02
iso $-C_4H_9OH$	1.63	1.70	1.04	1.04	1.02

A_2 only on the dimensions of the molecules. A_2 is extremely small and can be neglected in calculations.

The equation obtained by Conner and Smyth [58, 68] is somewhat different from equations (II.58) and (II.59):

$$\frac{\mu_s}{\mu_0} = 1 + 0.43A(\varepsilon_1 - 1) \qquad (II.60)$$

In the derivation of equation (II.49), the authors started from an ellipsoidal model of the molecule of the solute, i.e., they took into account the anisotropy of the polarizability of a polar molecule. According to (II.60), the factor 0.43 corresponds to $3/(\varepsilon_1 + 2)$ in the Higasi equation and to $1/\varepsilon_1$ in the Frank equation. Since the dielectric constant of frequently used solvents is generally between 1.9 and 2.5, the correction for the solvent effect obtained from the Conner—Smyth equation is 0.55-0.64 times the effect given by the Higasi equation and 0.52-0.65 times the figure obtained from the Frank equation, if the value of the constant in the latter is the same as A.

Equations (II.58), (II.59), and (II.60) are approximately equivalent and give comparatively close values of μ_s/μ_0. For example, the values of μ_s/μ_0 calculated by the Higasi, Frank, and Conner — Smyth equations are, respectively, 0.95, 0.96, and 0.97 (the experimental value being 0.88) and those for chlorobenzene are 0.91, 0.90, and 0.94 (the experimental value being 0.92). The equations considered can be used to calculate the solvent effect for various classes of compounds and, as follows from Table 6, which we have borrowed from Smyth's book [19], they give figures agreeing well with the experimental results. However, a fundamental defect of these equations is the complexity of the calculation of A, for which the dimensions and shape of the molecules must be known with sufficient accuracy.

It follows from the examples given that the Higasi and Conner — Smyth equations reflect correctly the direction and magnitude of the solvent effect. However, the Higasi — Frank — Conner — Smyth method by means of which the authors endeavor to calculate the additional moment arising through the polarization of the molecules of the solvent by the field of the polar molecule has given rise to a number of critical observations [41]. In particular, in the determination of the polarization induced at an arbitrary point of the solvent by the field of an ellipsoidal polar molecule and in the subsequent integration over some volume, the integral with respect to a finite volume must be a function of the shape of this volume. Consequently, the object (sample) investigated must have a definite shape. Furthermore, in the method considered no account is taken in integration of the field of the bound charges induced on the external surface of the sample [4].

Ross and Sack have investigated equation (II.51) for calculating the ratio μ_s/μ_0 [50]. The formula that they obtained has the following form:

$$\frac{\mu_s}{\mu_0} = \frac{3\varepsilon\,[1-(\varepsilon_\infty-1)\,D}{(\varepsilon+2)\,[\varepsilon+(\varepsilon_\infty-\varepsilon)\,D}\qquad\qquad (II.61)$$

Here ε_∞ is the optical dielectric constant of the solute. Since at infinite dilution ε tends to ε_∞, equation (II.61) becomes analogous to the Higasi equation. As has been mentioned above, D is the depolarizing factor. In calculation by formula (II.61), it is assumed that the dielectric constant and the refractive index are measured in dilute solution, even though the derivation of this formula does not depend on the Onsager approximation. Relation (II.61) has been verified for a number of substances, the calculated values of μ_s / μ_0 agreeing well with the experimental results.

The Higasi, Conner—Smyth, and Ross—Sack equations can be used successfully to find both the magnitude and the direction of the solvent effect. However, it must be borne in mind that none of these equations takes into account all the factors affecting the magnitude of the dipole moment of a polar molecule surrounded by molecules of a nonpolar solvent. The fact that in a number of cases corrections for the influence of the solvent may be comparable with the experimental errors must also be taken into account.

6. Determination of the Dipole Moment
by Means of the Stark Effect

The Stark effect is connected with the change in the energy level of atoms and molecules under the action of an external electric field and can be detected by a shift and a splitting of the spectral lines. This change in the spectra of a molecule can be used to determine the dipole moment.

The method of determining the dipole moment by means of the Stark effect is extremely accurate and demands that the applied field should possess a fairly high degree of uniformity.

The rotational level j corresponding to the total angular momentum is $(2j + 1)$-fold degenerate in the absence of an external field in accordance with the number of possible orientations of the vector of the angular momentum in space, $(2j + 1)$.

Since the energy of a molecule U with a constant dipole moment $\vec{\mu}$ in an external uniform electric field $\vec{E}$ depends on the mutual orientation of the vectors $\vec{\mu}$ and $\vec{E}$

$$U = -(\vec{\mu}\vec{E})$$

(II.62)

in the general case the energy of the states of the molecule corresponding to different orientations of the vector of the angular momentum will be different. The Stark effect in a uniform external electric field will be observed as a splitting of the line corresponding to the rotational level. The change in the magnitude of the splitting of the rotational lines in an external electric field permits the calculation of the constant dipole moment of the molecule.

Depending on the presence of components of the dipole moment directed along the vector of the total angular momentum, a Stark effect of the first order (splitting of the line proportional to the first degree of the strength of the external field $\vec{E}$) or a Stark effect of the second order (splitting proportional to the square of the field strength E^2), respectively, is observed. It is possible to obtain accurate formulas connecting the dipole moments with the magnitude of the splitting for the Stark effect in the case of linear molecules or a molecule of the symmetric spinning-top type.

A linear molecule has a second-order Stark effect. The splitting of the level may be expressed by the relation [69, 70]

$$\Delta\mathcal{E} = \frac{\mu^2}{2\hbar B}\left[\frac{j(j+1)-3M^2}{j(j+1)(2j-1)(2j+3)}\right]\vec{E^2} \tag{II.63}$$

where μ is the dipole moment of the molecule, $\hbar$ is Planck's constant divided by 2π, B is the moment of inertia of the molecule relative to an axis passing through the center of mass perpendicularly to the axis of the molecule, j is the quantum number of the angular momentum, and M is the quantum number of the projection of the vector of the angular momentum on the direction of the electric field $\vec{E}$.

In molecules of the symmetric-top type, a linear Stark effect is found [70]:

$$\Delta\mathcal{E} = \frac{2\mu KM}{j(j+1)(j+2)}\vec{E} \tag{II.64}$$

where K is the quantum number of the projection of the vector of the angular momentum on the axis of symmetry of the molecule.

If the external electric field E is known and the transition is reliably identified, from a measurement of the shift in the frequency of the line due to the Stark effect it is possible to determine the dipole moment of the molecule. A waveguide cell is usually used for these purposes. The external electric field is created by a plane electrode located within the waveguide. This arrangement of the plane electrode does not lead to an appreciable distortion of the microwave field and does not prevent the propagation of the microwaves, since the field is intersected by the electrode in a plane perpendicular to the direction of this field. It is essential to note that the electric field in such a waveguide has a considerable degree of nonuniformity only near the edges of the electrode and in

TABLE 7. Dipole Moments of Some Compounds
Determined by the Stark Method

Compound	μ, D	Compound	μ, D
Hydrogen cyanide	2.957	Acetone	2.8
Chloroacetylene	0.44	Isobutane	0.132
Nitromethane	3.46	Allylene	0.75
Methyl chloride	1.87	Propionaldehyde	2.46
Methyl bromide	1.797	Dimethyl sulfide	1.50
Methyl iodide	1.647	Formaldehyde	2.17
Methyl fluoride	1.79	Formyl fluoride	2.02
Hydrogen thiocyanate	1.72	Trifluoromethane	1.645
Chloroform	1.2	Methyl difluoroborane	1.67
Methyl phosphine	1.100	Methyl difluorosilane	2.11

the center it is parallel to the direction of the microwave
electric field. Consequently, in the calculation of the dipole moments the field may be considered uniform and equal in magnitude
to the field in the central part of the waveguide. The method of
measurement has been described in detail in the literature [69-71].

Accurate formulas for the Stark splitting in molecules of the
asymmetric spinning-top type are difficult to obtain. However, the
investigation of the Stark effect enables dipole moments to be calculated in this case, as well. When the direction of the dipole moment is unknown, the Stark effect is determined for several lines.
This makes it possible to calculate the value of the projections of
the dipole moment on the axis. This is a great advantage of this
method of determining dipole moments, since all the other methods
of measurement provide the possibility of determining only the
scalar magnitude of the dipole moment of a molecule. Determining
the dipole moment from the Stark effect makes it possible to measure small values of the dipole moment (of the order of 0.1-0.2 D)
with almost the same accuracy (about 0.2%) as large ones. It is
important that even a considerable contamination of the gases does
not affect the Stark method, since only those lines are selected for
measurement that belong to the molecule of interest. Table 7 gives
values of dipole moments of a number of simple compounds obtained by the Stark method. The error of the determination of the
moment does not exceed 1% of the measured value.

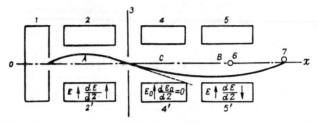

Fig. 15. Schematic diagram of the electric resonance method
of determining dipole moments: 1) Source of the molecular
beams; 2, 2') electrodes creating an electric field in region A
in the direction 2'→2 with a gradient in the direction 2' → 2;
3) collimator slit; 4, 4') electrodes creating a uniform electric
field in the direction 4' → 4 in region C, in which a variable
electric field is also created; 5, 5') electrodes creating an electric field in region B in the direction 5' → 5 with a gradient in
the direction 5 → 5'; 6) barrier filament eliminating molecules
with small effective dipole moments; 7) detector.

The high sensitivity of the method under consideration enables us to measure dipole moments in excited vibrational states
of molecules and to find values of the moments for each individual
vibrational state. For example, it has been established that the dipole moment of a molecule in the first excited state corresponding
to bending vibrations is 0.700 D, which is 1.2% less than the value
for the lower state (0.709 D) [69].

Unfortunately, the Stark method for determining dipole moments is limited by the necessity of studying in the gas phase substances possessing a fairly high vapor pressure.

7. Electric Resonance Method
of Determining Dipole Moments

Before we discuss the electric resonance method, we shall
dwell very briefly on the determination of dipole moments by the
molecular beam method.

The method of deflecting molecular beams in an electric field
can be used to determine dipole moments. A molecule moving
along the direction of the axis OX (Fig. 15) in a nonuniform electric
field E_0 which, like its gradient, is perpendicular to OX, is subject

to the action of a force $\mu_{Z_{eff}}$ dE_0/dz, causing a deviation of the beam in the direction of the Z axis. Here $\mu_{Z_{eff}}$ is the component of the dipole moment of the molecule present in the electric field E_0 in the direction of the Z axis. By measuring the displacement of the beam it is possible to calculate the dipole moments of linear molecules of alkali-metal halides, etc. The low accuracy of the method is due to the presence of molecules in different states in the beam, to a distribution of molecules with respect to velocities, and to the inaccuracy of measuring the gradient of the electric field [72-73].

These effects of the molecular beam method are eliminated when the electric resonance method of determining dipole moments is used. The basic principles of this method are illustrated in Fig. 15.

A beam of molecules from the source 1 is directed along the axis of the system. The molecules pass through two regions A and B of a nonuniform field with strictly opposed gradients dE/dZ and a region with a uniform field C. The construction of the apparatus is such that the molecules of the beam present in a single definite rotational state converge after passing through the regions A, B, and C (if no change in their orientation takes place en route) at a single point, at which a particle detector is placed. A small high-frequency field acting in the region C with a uniform field $\vec{E}_0$ induces (when the frequency is chosen correctly) transitions of the molecules into states with different orientations of the dipole moment, causing changes in the trajectory in region B; as a result, the molecule does not meet the detector. By changing the frequency of the field and measuring the change in the particle flux incident on the detector, the frequency at which resonance takes place is determined. Thus, in the resonance method transitions between levels of the molecules in a uniform static field $\vec{E}_0$ corresponding to different orientations of the dipole moment are studied. The dipole moments only of linear molecules are determined by the electric resonance method. The formula providing the possibility of calculating the dipole moment of a linear molecule μ from the measured frequency ν at the transition $(1,0) \rightarrow (1, \pm 1)$ [72] can be given in the form

$$hv = \frac{3\,(\mu E_0)^2}{20\,(\hbar^2/2B)} - \frac{0.00959\,(\mu E_0)^4}{(\hbar/2B)^3} \qquad \text{(II.65)}$$

where the magnitude B has the same meaning as in formula (II.63).

The electric resonance method possesses a high accuracy and makes it possible to observe transitions between energy levels connected with definite rotational states of polar molecules, even in those cases where the number of molecules in these states is 10^{-4} of the total beam. However, the complexity of the apparatus and serious experimental difficulties limit the scope of its application.

8. Determination of Dipole Moments
of Liquids by Measuring Dielectric
Losses in the Microwave Region

In liquids, the large number of collisions between neighboring molecules interferes with the observation of discrete rotational states. However, polar liquids have a wide absorption band in the microwave frequency range which arises as a consequence of the orientation of the molecules in an irradiation field with an increase in the vector of the electric field $\vec{E}$ and the return of the molecules to the state of thermal equilibrium when the vector decreases to zero [69]. This wide region of absorption has maxima at the frequencies

$$v_0 = \frac{1}{2\pi\tau} \qquad \text{(II.66)}$$

where τ is the mean relaxation time for the reorientation of the molecular dipoles. By measuring the frequency of the absorption maximum it is possible to determine τ.

The absorption is generally defined in terms of tan δ, where δ is the loss angle. According to Debye [15, 16], for dilute solutions of polar substances in dipole–free solvents

$$\tan \delta = \frac{(\varepsilon + 2)^2}{\varepsilon} \cdot \frac{4\pi\mu^2 C N v}{27kTv_0\left[1+\left(\dfrac{v}{v_0}\right)^2\right]} \qquad \text{(II.67)}$$

where ε is the static dielectric constant of the solution, μ is the dipole moment of a molecule of the polar substance, v is the irradiation frequency, $v_0 = 1/2\pi\tau$ is the frequency of the absorption maximum, and C is the concentration.

By measuring the tangent of the loss angle at three or more different frequencies and using equation (II.56) ε, μ, and ν_0 can be calculated. It is also possible to carry out the measurements at constant frequency and variable temperature and obtain the data necessary for determining the dipole moment, the average relaxation time, and the dielectric constant.

Literature Cited

1. L. D. Landau and E. M. Livshits, Field Theory, Fizmatgiz, Moscow (1960), p. 120.
2. I. E. Tamm, Principles of Electricity Theory, Gostekhizdat (1956).
3. R. S. Mulliken, J. Chim. Phys., 46:497 (1949).
4. C. Coulson, Valence, Oxford University Press, 2nd Edn. (1961).
5. R. S. Mulliken, C. A. Rieke, D. Orloff, and H. Orloff, J. Chem. Phys., 17:1248 (1949).
6. R. S. Mulliken, J. Chim. Phys., 3:573 (1935).
7. C. A. Coulson, Trans. Faraday Soc., 38:433 (1942).
8. C. A. Coulson and M. T. Rogers, J. Chem. Phys., 35:593 (1961).
9. A. K. F. Duncan and J. A. Pople, Trans. Faraday Soc., 21:851 (1953).
10. K. Ruedenberg, "The physical nature of the chemical bond," Rev. Mod. Phys., 34:326-376 (1962).
11. C. P. Smyth, The Dielectric Constant and the Structure of Molecules, Chemical Catalog Co. Inc., New York (1931).
12. O. A. Osipov and V. I. Minkin, Handbook of Dipole Moments, Izd. Vysshaya shkola (1965).
13. A. A. Maryott and F. Buckley, Table of Dielectric Constants and Electric Dipole Moments of Substances in Gaseous State, NBS Circular 573, Washington (1953).
14. Landolt-Börnstein, Bd. 1, T.3, 14,206, Berlin-Göttingen-Heidelberg, 1951.
15. P. Debye, Polare Molekeln, Leipzig (1929) [English translation, The Chemical Catalog Co., New York (1929)].
16. P. Debye and H. Sack, Die Theorie der elektrischen Eigenschaften von Molekülen [Russian translation] ONTI, Moscow (1936).
17. C. T. Zahn, Phys. Rev., 27:445 (1926).
18. R. Sänger, O. Steiger, and K. Gächter, Helv. Phys. Acta, 5:136 (1929).
19. C. P. Smyth, Dielectric Behavior and Structure, McGraw-Hill, New York (1955).
20. A. Weissberger, Ed., Physical Methods in Organic Chemistry [Vol. I of Technique of Organic Chemistry, Interscience, New York, 2nd Edn. (1949)].
21. K. L. Wolf, Z. Phys. Chem., B2:39 (1929).
22. A. Weissberger, Ed., Physical Methods in Organic Chemistry [Vol. I of Technique of Organic Chemistry, Interscience, New York, 1st Edn. (1946)].
23. B. V. Ioffe, Refractometric Methods in Chemistry, GNTI, (1960).
24. G. Hedestrand, Z. Phys. Chem., B2:428 (1929).
25. H. de Vries Robles, Rec. Trav. Chim., 58:111 (1939).
26. R. J. W. LeFevre and H. Vine, J. Chem. Soc., 1937:1805.
27. R. J. W. LeFevre, Trans. Faraday Soc., 46:1 (1950).

28. I.F. Halverstadt and W.D. Kumler, J. Am. Chem. Soc., 64:2988 (1942).
29. T. Fujita, J. Am. Chem. Soc., 79:2471 (1957).
30. F.R. Gross, J. Chem. Soc., 1937:1915.
31. E.A. Guggenheim, Trans. Faraday Soc., 45:714 (1949).
32. E. Guggenheim and J. Prue, Physicochemical Calculation, North-Holland Publishing Co., Amsterdam (1955).
33. E.A. Guggenheim, Trans. Faraday Soc., 47:573 (1951).
34. S.R. Palit, J. Am. Chem. Soc., 74:3952 (1952).
35. J.W. Smith, Trans. Faraday Soc., 46:394 (1950).
36. K. Higasi, Bull. Inst. Phys. Chem. Res., 22:805 (1943).
37. K. Bal and S.C. Srivastava, J. Chem. Phys., 32:663 (1960).
38. K. Bal and S.C. Srivastava, J. Chem. Phys., 27:835 (1957).
39. S.C. Srivastava and P. Charandas, J. Chem. Phys., 30:816 (1959).
40. K.B. Everard, R.A.W. Hill, and L.E. Sutton, Trans. Faraday Soc., 46:417 (1950).
41. W. Brown, Dielectrics, [Russian translation] IL, Moscow (1961).
42. H. Fröhlich, The Theory of Dielectrics, Clarendon Press, Oxford (1949).
43. E.A. Guggenheim, Proc. Phys. Soc., 68B:186 (1955).
44. C.J.W. Böttcher, Theory of Electric Polarisation, Elsevier, Amsterdam (1952).
45. J.W. Smith, Electric Dipole Moments, Butterworths, London (1955).
46. W.D. Kumler, N. Pearson, and F.V. Brutcher, J. Am. Chem. Soc., 83:2711 (1961).
47. C.M. Lee and W.D. Kumler, J. Am. Chem. Soc., 83:4586 (1961).
48. W.D. Kumler, A. Lewis, and S. Meinwald, J. Am. Chem. Soc., 83:4591 (1961).
49. T.G. Scholte, Rec. Trav. Chim., 70:50 (1951).
50. J.G. Ross and R.A. Sack, Proc. Phys. Soc., B64:619, 620 (1951).
51. A. Weissberger, Ed., Physical Methods in Organic Chemistry [Vol. I of Technique of Organic Chemistry, Interscience, New York, 2nd Edn. (1949)].
52. C.J.W. Böttcher, Physica, 6:59 (1939).
53. F.H. Müller, Phys. Z., 34:689 (1933); Trans. Faraday Soc., 30:731 (1937).
54. F.H. Müller and P. Mortier, Phys. Z., 36:371 (1935).
55. F.H. Müller, Phys. Z., 38:283 (1937).
56. J.G. Ross and R.A. Sack, Proc. Phys. Soc., B63:893 (1950).
57. J.R. Partington, An Advanced Treatise on Physical Chemistry, Vol. 5, Longmans, Green, and Co., London (1954).
58. W.P. Conner, R.P. Clark, and C.P. Smyth, J. Am. Chem. Soc., 64:1379 (1942).
59. S. Sugden, Nature, 133:415 (1934).
60. C.G. LeFevre and R.J.W. LeFevre, J. Chem. Soc., 1935:1747.
61. H.O. Jenkins, Nature, 129:106 (1934).
62. E.A. Guggenheim, Nature, 137:459 (1936).
63. M.A. Govinda Rau, Proc. Indian Acad. Sci., 1:489 (1934).
64. F.H. Müller, Phys. Rev., 50:547 (1936).
65. H.O. Jenkins and S.H. Bauer, J. Am. Chem. Soc., 58:2435 (1934).
66. K. Higasi, Sci. Papers Inst. Phys. Chem. Res., 28:284 (1936).
67. F.C. Frank, Proc. Roy. Soc., 152A:171 (1935).
68. R.H. Wiswall and C.P. Smyth, J. Chem. Phys., 9:356 (1941).
69. C. Townes and A. Schawlow, Microwave Spectroscopy, McGraw-Hill, London (1955).

70. D. Ingram, Spectroscopy and Radio and Microwave Frequencies, Butterworths, London, First Edn. (1955).
71. W. Gordy, W. Smith, and R. F. Trambarulo, Radiospectroscopy, Wiley, New York (1953).
72. K. F. Smith, Molecular Beams, Wiley, New York (1955).
73. N. Ramsey, Molecular Beams, Clarendon Press, Oxford (1956).

Chapter III

Calculations in the Dipole Moment Method

1. Bond and Group Moments

The investigation of the majority of structural problems using the dipole moments of molecules is based on a comparison of the experimental values of the moments with those calculated by the additive method. The basic idea of the latter was suggested by Thomson [1] in 1923; he showed that the dipole moment of poly-atomic molecules $\vec{\mu}$ can be considered as the resultant of the vectorial combination of the moments belonging to the individual bonds. For a molecule containing n bonds of different types,

$$\vec{\mu} = \sum_{i=1}^{n} \vec{\mu}_i \qquad (\text{III.1})$$

where $\vec{\mu}_i$ is the moment characterizing the i-th bond.

Thus, if the geometry of the molecule is known or can be assumed, calculation of the dipole moment is carried out by the usual rules of vectorial combination of the vectors of the bond moments. The determination of the absolute magnitudes and directions of the bond moments is an extremely difficult task the solution of which largely determines the value and reliability of the structural information obtained by the dipole moment method. Let us consider various methods of approaching the calculation of bond moments.

Theoretical and Semiempirical Evaluations

of Bond Moments

The study of the nature of the dipole moment of a diatomic molecule carried out by Coulson [2] and Mulliken [3] (Chapter II) shows that the polarity of the bond is determined not only by the effective charges localized on the atoms forming the bond but also by such contributions as the atomic (hybridization) and homopolar dipoles and also the dipole due to the presence of unshared pairs of electrons in hybrid orbitals. An accurate calculation of the influence of all factors mentioned is difficult at the present time and theoretical calculations of dipole moments of individual bonds are, in the main, of purely qualitative value.

Particular attention has been devoted to calculations of the moment of the $C-H$ bond, as the key magnitude in the analysis of the dipole moments of organic compounds.

Coulson [2], using the wave functions of the self-consistent field of methane, derived for the dipole moment of the $C-H$ bond a value of 0.4 D with the polarity C^+H^-. A similar result using a more simplified quantum mechanical treatment was obtained by Hartmann [4]. A review of other early investigations on the calculation of the moment of the $C-H$ bond was made by Gent [5]. Taking for the moment of the $C-H$ bond in methane the value put forward by Coulson, Gent showed that an increase in the s nature of the hybrid orbital of the carbon involved in the bond with the hydrogen atom must lead to a reversal of the direction of the moment of the $C-H$ bond in acetylene as compared with the moment of this bond in methane. According to Gent [5] and Walsch [6], the moment of the $C-H$ bond in ethylene is close to zero, and its polarity in acetylene (C^-H^+) is opposite to that in methane [7]. The main value of these data consists in the conclusion to which they lead that the magnitude of the dipole moment of a bond depends essentially on the nature of the hybridization of the atoms forming it.

The values of the moments of $C-H$ bonds obtained from quantum mechanical calculations depend very greatly on the approximations used and vary, for example, for the $C-H$ bond in methane from 0.6 (C^-H^+) [8] to -2.8 D (C^+H^-) [9]. Even in the most correct calculations, a whole series of factors was left out of account (secondary dipole moments induced in the system of

electrons not participating directly in the bonding, the inductive influence of different carbon-hydrogen bonds, spatial interactions,
etc.). Each of them could seriously affect the final result.

A particularly important conclusion proceeding from the theoretical consideration of bond moments is the idea of the moment
contributed by the unshared pair of electrons with nonbonding hybrid orbitals of such atoms as N, O, Cl, and others. The direction
of the moment of the unshared pair is determined unambiguously in
such a way that the negative end of the dipole is localized on the
axis of the hybrid orbital in the "center of gravity" of its electron
cloud. As a result, for any states of hybridization of the unpaired
electrons, apart from the sp state, the direction of the vector of
the moment of the unshared pair of electrons does not coincide
with the direction of the vectors of the moments of the bonds
formed by the central atom and, consequently, the moment of the
unshared pair cannot be included in a simple manner in the bond
moment and must be taken into account as a separate vector in calculations using formula (III.1).

Many attempts have been made to calculate the moments of unshared electron pairs of atoms in various states of hybridization.
In those calculations in which Slater orbitals were used as wave
functions, very high values of the moments of the unshared electron
pairs were obtained: 3.19 D for the moment of the electron pair of
the oxygen of water [10], from 3.50 to 3.7 D for the moment of the
electron pairs of nitrogen in ammonia [10, 11], pyridine [12], and
tetrazole [13], 4.86 D for the moment of the electron pair of phosphorus in phosphine [11], etc. These magnitudes considerably exceed the majority of bond moments. However, in rigorous calculations using the functions of the self-consistent field, lower values
for the moments of unshared electron pairs are found: for the oxygen of water 1.69 [14, 15], for the nitrogen of ammonia 1.45, and
for the nitrogen of trimethylamine 0.64 D [16].

The dipole of an unshared electron pair is a magnitude difficult to measure experimentally. However, the few attempts that
have been made in this direction indicate comparatively small values of the moments of unshared electron pairs (in comparison
with the calculated values). Thus, having determined the π component of the dipole moment of pyrimidine from the data of the NMR
spectrum and then having deducted the magnitude of the polar mo-

ment of its σ-component, Seiffert et al. [17] give a value 0.8 D for the moment of the unshared sp^2 pair of nitrogen.

An extremely attractive prospect of determining the moment of an unshared pair is based on a comparison of the atomic coordinates obtained from X-ray and neutron-beam measurements [18]. In the latter case, the measurements give the position of the atomic nuclei and in the former case the centers of gravity of the electron clouds of the atoms. For an atom with a hybrid unshared electron pair, the center of gravity of the electron cloud is displaced, as shown above (Fig. 9) by a distance $\bar{x}$ from the nucleus. By determining $\bar{x}$ as the difference in the vectors corresponding to an atom with an unshared pair found by X-ray and neutron-beam measurements, it is not difficult to calculate the moment of the unshared pair of electrons as $\mu_a = ze\bar{x}$, where z is the charge on the nucleus of the neutral atom.

At the present time, a small number of compounds is known for which both X-ray and neutron-beam measurements have been carried out. These compounds include ammonia and hexamethyl-enetetramine. In both cases $\bar{x}$ is approximately 0.025 Å, which corresponds to a moment of the unshared sp^3 pair of nitrogen elec-trons of 1.0 D.

The figures given show that, although the moments of the un-shared pairs must also be taken into account as far as possible in vector calculations of dipole moments, their value is by no means as large as is sometimes assumed. On the other hand, for the ma-jority of bonds the atomic and homopolar components of the bond moment have the opposite direction and compensate one another to a considerable degree. Considering also that atomic dipoles de-pend comparatively little on the type of atoms forming the bond and that in molecules with a tetrahedral or trigonal arrangement of the hybrid bonds they must substantially counterbalance one another [19], there are grounds for attempting to undertake the construction of a scale of bond moments starting from the effective charges of the atoms.

In these considerations, two approaches have been used which are based on the use of ideas on the electronegativities of atoms and groups and on semiempirical calculations by the local-ized molecular orbital method. In the first of them, the dipole moment of a bond $A-B$ is defined as

$$\mu_{A-B} = 4.8ri \tag{III.2}$$

where r is the length and i is the ionicity of the bond, i.e., a magnitude characterizing the excess of the electronic charge on atom B in the $A-B$ bond. Thus, if the $A-B$ bond is regarded as a localized molecular orbital $a\Psi_A + b\Psi_B$, on the assumption that the atomic orbitals are orthogonal, the charge on atom B is $1-2b^2$.

The value of i for various bonds can be calculated from the functional dependence of the ionicity of the bonds on the electronegativity X of the atoms forming them. The best-known relation is that of Pauling [20]

$$i = 1 - e^{-0.25\,(X_A - X_B)^2} \tag{III.3}$$

In Batsanov's book [21] there is a detailed review of the numerous equations connecting the electronegativities of atoms with the ionicities of the bonds constructed from them.

Recently, Pauling's concept of electronegativity has been supplemented by ideas on the dependence of the electronegativities of atoms on their chemical environment and hybridization [21, 22] and also by new definitions of electronegativity, for example, in terms of orbital and bond electronegativities [23]. New expressions for calculating bond ionicities have been formulated.

An original approach to the calculation of dipole moments of molecules upon which a refined concept of electronegativity has been based has been proposed by Bykov [24].

In practical calculations of dipole moments of complex molecules, the bond moments found by means of relations of type (III.2) and (III.3) are not usually used. However, cases can also be found in which the dipole moments of molecules containing one or several bonds with unknown bond moments must be calculated. In the absence of other methods for their calculation, or if an approximate evaluation is adequate, formulas (III.2) and (III.3) give a very simple and convenient method for finding a bond moment. A determination of the moment of the $Si-N$ bond in compounds of the type $(CH_3)_3SiNHAlk$ [25], the value of which was difficult to find by any other method because of the absence of compounds in which the $Si-N$ bond is the only polar bond, can be used as an example.

Del Re [19] calculated the values of some bond moments after having obtained the charge distribution for a number of saturated molecules by using an empirical variant of the method of localized molecular orbitals (see § 4). These calculations took into account induction effects reflected in values of the bond moments in different compounds. However, they enabled a range of variations of the bond moments in molecules of saturated compounds to be established:

Bond	Bond moment	Bond	Bond moment
H—C	0.20—0.28	C—O	0.93—1.14
C—N	0.47—0.62	H—O	1.44—1.60
H—N	1.08—1.21	C—Cl	0.83—1.65

Note: The atom shown on the left is the positive pole of the dipole.

The nature of the polarity of a bond is determined by the signs of the effective charges of the atoms. On comparing the values calculated by Del Re with those of Table 8 obtained from experimental values of the dipole moments, good agreement of the corresponding bond moments is found. This permits us to regard the method of calculating bond moments in which atomic and homopolar dipoles and also dipoles of unshared electron pairs are neglected as fairly satisfactory, although an extremely coarse approximation. Some theoretical justification of this and the approximation of the point distribution of charges in the molecule that is generally used for calculations of dipole moments has been given by Lipscomb et al. [26].

Calculation of Bond Moments from the Intensities

of Vibrational Absorption Bands

The measurement of the intensities of vibrational absorption bands is the only experimental method of evaluating bond moments that is not connected with the direct determination of the dipole moments of molecules. Consequently, the results obtained by this method are of particular interest.

The integrated intensity A of an absorption band corresponding to some fundamental vibration is governed by the relation

$$A = \frac{N\pi}{3c^2} \left(\frac{\partial \vec{\mu}}{\partial Q} \right)^2 \tag{III.4}$$

where $\partial\vec{\mu}/\partial Q$ is the component of the vector of the dipole moment along the normal coordinate Q. Consequently, the absolute values of $\partial\mu/\partial Q$, but not the sign, can be determined from experimental data with accuracy, and for some symmetrical molecules they can also be characterized as vectors. In the general case, the direction of the vector $\partial\vec{\mu}/\partial Q$ is determined by measuring the polarization of the vibrational absorption bands.

The magnitudes of $\partial\vec{\mu}/\partial Q$ can be represented as functions of the natural vibrational coordinates Q and the dipole moments of the individual bonds μ_i, which are also capable of determination. Methods of solving this problem have been considered in detail, for example, in reviews [27, 28]. The most general method is based on the valence-optical theory [29].

In several previous investigations, substantial differences were reported in the magnitudes of the bond moment determined from the intensities of the bands of the IR spectra due to different types of vibrations. Thus, the moment of the C—H bond in ethylene was estimated on the basis of the intensity of the nonplanar vibrations of C—H bonds as 0.68 D with a polarity of the C^-H^+ type. At the same time, an analysis of two types of planar vibrations gave values from +0.26 to −0.26 D [30]. Similar inconsistencies led to the formulation of ideas according to which the results of calculations from the intensities of the bands of the IR spectra are vibrational or dynamic bond moments the nature of which is different from the static moments considered above [27, 31, 32]. The unsoundness of this assumption was stressed by Gribov [28]. Sverdlov [33] showed that the differences in the magnitudes of the bond moments that have been mentioned were due to impermissible simplifications of the equations connecting the dipole moments of the bonds with the intensities and that these differences disappeared with a more rigorous approach.

At the present time, only a few bond moments, especially the moments of C—H bonds, have been calculated fairly reliably. The results obtained satisfactorily confirm the predictions of the theory of the fundamental dependence of these magnitudes on the nature of the hybridization of the carbon atom.

Cole and Michell [34], by analyzing the intensities of the bands of the nonplanar deformation vibrations of benzene and 15 of its methyl and halogen derivatives, found a value of the $C_{sp^2}^- - H^+$ moment of 0.64 ± 0.01 D. Extremely similar values in the range from 0.63 to 0.70 D have been obtained by other authors [28, 35]. The moment of the $C_{sp^2}^- - H^+$ bond in ethylene has been estimated at 0.7 D [28, 36].

The $C_{sp}^- - H^+$ moment has been determined from the intensities of the vibrational bands in the spectra of hydrogen cyanide and acetylene and is, according to the results of various authors, 1.05–1.1 D [27, 28].

Quite good agreement has also been obtained for the magnitudes of the moments of $C_{sp^3} - H$ bonds. The main difficulty here is the determination of the nature of the polarity of this bond. Hiller and Straley [37], using data on the intensities of the vibrational bands of deuterated methanes give a value of 0.33 D, tending to assume a polarity of the type $C^+ H^-$. For ethane and dimethylacetylene, Gribov [28] found a value of the moment of the $C_{sp^3} - H$ bond of 0.28 D with the polarity $C^- H^+$. For substituted methanes, the following $C_{sp^3} - H$ moments (in D) have been obtained:

CH_3I 0.20	CH_3F . . . . . 0.413 [38]	
CH_3Br 0.23	CH_3NO_2 . . . . 0.30 [39]	
CH_3Cl . . . . 0.25 [28]		

The figures given for the moments of $C - H$ bonds agree with the results obtained from measurements of the IR dispersion of methane [40] ($C_{sp^3}^- - H^+ = 0.31$ D), ethylene [41] ($C_{sp^2}^- - H^+ = 0.63$ D), and acetylene [42] ($C_{sp}^- - H^+ = 1.05$ D).

The increase in the bond moments in the sequence $H - C_{sp^3}$, $H - C_{sp^2}$, $H - C_{sp}$ corresponds to the accepted ideas according to which an increase in the s nature of a hybrid orbital of carbon increases its effective electronegativity, since the 2s orbital is centered closer to the nucleus than the 2p orbital [43].

There are considerably less detailed results based on the intensities of vibrational bands for the moments of other bonds. This is due mainly to the extreme laboriousness of the calculations of bond moments from vibrational intensities and also to the fact that the possibilities of the methods of calculation are fundamentally limited by the dimensions and degree of symmetry of the molecule under study. In view of this, the main source of information on bond moments is the analysis of dipole moments of molecules.

Calculation of Bond and Group Moments

from the Dipole Moments of Molecules

B o n d   M o m e n t s .   Relation (III.1) may obviously be used not only to calculate the dipole moment of a molecule from the moments of the individual bonds but also to solve the inverse problem: to resolve the total moment into its components.

Thus, the moments of the $H-O$ and $H-N$ bonds can be calculated from a knowledge of the dipole moments of water and ammonia, respectively

$$2\mu_{O-H} \cos 52.5° = 1.84\ D$$
$$\mu_{O-H} = 1.51\ D$$

$$3\mu_{N-H} \cos 68° = 1.46\ D$$
$$\mu_{N-H} = 1.31\ D$$

With such a method of calculation it is possible to obtain only the absolute magnitude but not the direction of the vectors of the bond moments. It is generally assumed that the vector of a bond moment is directed towards the more electronegative atom of the bond concerned. The moment contributed to the atomic dipole by an unshared electron pair is in fact resolved into components collinear with the bonds, and the bond moments calculated in this way include the contribution of the dipoles of the unshared electrons.

The most complex problem is to calculate the moment of the $C-H$ bond. On the assumption that the $C-H$ moment in the phenyl nucleus and the moment of the CH_3-C_{phenyl} bond of toluene were equal to zero, Eucken and Meyer [44] concluded that the dipole moment of toluene (0.37 D) was due to the vectors of the moments of the three $C-H$ bonds of the methyl group and therefore the moment of the $C-H$ bond, assuming a tetrahedral HCH angle, was 0.37 D.* Both of Eucken and Meyer's initial assumptions proved to be erroneous, but the final result closely agreed with the results of deter-

---

* In a tetrahedral molecule CH_3X with angles of 109.5° between the bonds, each $C-H$ bond has a component $\mu_{C-H}\cos(180° - 109.5°) = \mu_{C-H}/3$ directed along the $C-X$ axis. Consequently, the total moment of the three $C-H$ bonds directed along the $C-X$ axis is μ_{C-H}.

TABLE 8. Bond Dipole Moments*

Bond	Bond moments, D		Bond	Bond moments, D	
	$\mu_{C^- - H^+} =$ $= 0.4D$	$\mu_{C^+ - H^-} =$ $= 0.4D$		$\mu_{C^- - H^+} =$ $= 0.4D$	$\mu_{C^+ - H^-} =$ $= 0.4D$
C—F	1.39	2.19	C=O	2.4	3.2
C—Cl	1.47	2.27	C=S	2.0 [50]	—
C—Br	1.42	2.22	C≡N	3.1	3.9
C—I	1.25	2.05	H—O	1.51	1.51
C—N	0.45	1.26	H—N	1.31	1.31
C—O (alcohols)	0.7	1.9	H—S	0.7	0.7
C—O (ethers)	0.7	1.5	Si— C	1.2 [51]	—
C—S	0.9	1.6	Si—H	1.0 [52]	1.0
C—Se	0.7	1.5	Si—N	1.55 [25]	1.55
C=N	1.4	—			

---

* The values of the bond moments were calculated from the most accurate
data [48] on the dipole moments of the methyl derivatives (CH₃Hal, CH₃OH,·
(CH₃)₃N, (CH₃)₂CO, etc.), under standard conditions (benzene, 25°C). Point
geometry of the molecules assumed [49]. The atoms shown on the left of the
bond are the positive pole of the dipole.

minations of the moment of the C —H bond from IR spectral data
(see above) and was used to construct a scale of bond moments
[45-47]. In this, the polarity of the C —H bond was taken as C^-H^+,
although the contrary possibility C^+H^- was also not excluded.

The choice of the direction of the vector of the moment of
C —H bond has a considerable effect on the values of the bond moments calculated from the dipole moments of organic compounds.
For example, on the assumption that $C^-H^+ = 0.4$ D, the bond moment of C —Cl with the polarity C^+Cl^- is

$$\mu_{C^+ - Cl^-} = \mu_{CH_3 - Cl} - \mu_{CH_3 - C} = \mu_{CH_3 - Cl} - \mu_{C - H} = 1.87 - 0.4 = 1.47 \ D$$

and on the assumption that $C^+H^- = 0.4$ D

$$\mu_{C^+ - Cl^-} = 1.87 + 0.4 = 2.27D$$

Table 8 gives the bond moments that are most important for
a study of the dipole moments of organic compounds calculated for
the two directions of the vector of the C —H bond. These figures
were obtained for bonds in saturated compounds. The moments of
the same bonds in aromatic compounds will differ from the values
given in Table 8, since the moment of the C_{sp^2} —H bond differs from
the moment of the C_{sp^3} —H bond in aliphatic compounds. Thus, taking the value 0.7 D for the moment of the $C_{sp^2}^-$ —H$^+$ moment [36],

TABLE 9. Dipole Moments of Carbon—Carbon Bonds

Bond	Bond moment, D	Compound from the dipole moment of which the bond moment was calculated
$C^+_{sp^3}-C^-_{sp^2}$	0.69 0.67	Toluene (0.37 D) Propylene (0.35 D)
$C^+_{sp^3}-C^-_{sp}$	1.48	Propyne (0.75 D)
$C^+_{sp^2}-C^-_{sp}$	1.15	Phenylacetylene (0.73 D)

we obtain for the moment of the $C^+_{sp^2}$ —Cl^- bond in chlorobenzene:

$$\mu_{C-Cl} = \mu_{C_6H_5-Cl} - \mu_{C-H} = 1.59 - 0.7 = 0.89\ D$$

which differs considerably from the C_{sp^3} —Cl moment calculated above.

Simple calculations show that the assumption of a relationship between the moment of the C —H bond and the hybridization of the carbon atom leads to the conclusion that carbon—carbon bonds formed by atoms with different hybridizations are polar. Petro [41], assuming the following values of the bond moments from the results of IR measurements: $C^-_{sp^3}$ —H^+ = 0.31, $C^-_{sp^2}$ —H^+ = 0.63, and C^-_{sp} —H^+ = 1.05 D, calculated the moments of carbon—carbon bonds given in Table 9.

Group Moments. In many cases, in the calculation of molecular dipole moments it is more convenient to use not bond moments but moments of individual groups of atoms. For example, in an analysis of the dipole moments of aromatic nitro compounds it is not necessary to calculate individually the moments of the C—N, N=O, and N → O bonds, but it is advantageous to use the resultant magnitude of the group moment μ_{NO_2}

In this case, μ_{NO_2} is taken as equal to the dipole moment of benzene and, consequently is in fact the sum of the moment of the nitro group and of a C —H bond in the p-position. This is very

convenient since it permits the omission of the moments of the
C$-$H bonds in explicit form from the calculation. In actual fact,
the dipole moment of m-dinitrobenzene, for example, can be calculated simply as the vectorial sum of the two μ_{NO_2} moments.

A single atomic group has different group moments according to whether it is present in an aliphatic or an aromatic molecule. Thus, the dipole moment of acetonitrile in benzene (25°C) is
3.47 D, while that of benzonitrile under the same conditions is
4.05 D. This difference is explained by the difference in the moments of the $C_{sp3}-H$ and $C_{sp2}-H$ bonds, which make a contribution
to the dipole moments of, respectively, acetonitrile and benzonitrile, by the different hybridizations of the carbon atoms to which
the nitrile group is attached, and by the effect of the conjugation of
the nitrile group with the aromatic nucleus in benzonitrile. It is
difficult to take all these factors into account accurately. Consequently, it is customary to distinguish values of group moments in
aromatic and aliphatic compounds.

The value of a group moment in the aromatic series is taken
as the value of the dipole moment of the monosubstituted benzene,
and in the aliphatic series that of the corresponding monosubstituted methane. The direction of the vector of the group moment can
be established by comparing the experimental values for p-substituted chloro- or nitrobenzene with the values calculated for the
two possible orientations of the vector moment of the substituting
group [53]. Thus, from a comparison of the experimental value of
the dipole moment of p-chlorophenyl azide of 0.30 D with that calculated from the dipole moments of chlorobenzene and phenyl azide

$$\bar{N}{=}\overset{+}{N}{=}N{\leftarrow}\langle\ \rangle{\rightarrow}Cl \qquad \bar{N}{=}\overset{+}{N}{=}N{\rightarrow}\langle\ \rangle{\rightarrow}Cl$$

$$\mu_{calc} = 1.59 - 1.44 = 0.15\ D \qquad \mu_{calc} = 1.59 + 1.44 = 3.03\ D$$

it is clear that the vector of the group moment of the azido group
is directed from the ring to the azido group.

For those groups of atoms X that have no axis of symmetry
coinciding with the line of the C$-$X bond (for example, $-C{\Big\langle}^{O}_{H}$,
$-C{\Big\langle}^{O}_{OAlk}$, $-O{\Big\langle}^{H}$, etc.) the group moment is fixed at some angle
θ to the bond mentioned. Such groups are usually called irregular,
in contrast to the regular groups $-NO_2$, $-CN$, $-C{\equiv}CH$, etc. The

TABLE 10. Group Moments (Benzene, 25°C)*

Group X	$\mu_{C_6H_5-X},\ D$	$\theta°,$	$\mu_{CH_3-X},\ D$	$\theta°,$
CH_3	0.37	0	0	0
CF_3	−2.54	0	−2.32	0
CCl_3	−2.04	0	−1.57	0
CN	−4.05	0	−3.47	0
CHO	−2.96	146	−2.49	125
$COCH_3$	−2.96	132	−2.75	120
COOH	—	—	−1.63	106
$COOCH_3$	−1.83	110	—	—
$COOC_2H_5$	−1.9	118	1.8	89
F	−1.47	0	−1.79	0
Cl	−1.59	0	−1.87	0
Br	−1.57	0	−1.82	0
I	−1.40	0	−1.65	0
OH	1.55	90	—	—
OCH_3	1.28	72	1.28	124
OCF_3	−2.36	160	—	—
$OCOCH_3$	1.69	66	—	—
NH_2	1.53	48,5	1.46	91
$NHCH_3$	1.71	40	—	—
$N(CH_3)_2$	1.58	30	0.86	109
$NHCOCH_3$	−3.69	100	—	—
N_3	−1.44	0	—	—
NO	−3.09	149	—	—
NO_2	−4.01	0	−3.10	0
SH	−1.22	135	−1.55	—
SCH_3	1.34	77,5	—	—
SCF_3	−2.50	156	—	—
SF_5	−3.44	0	—	—
SO_2CH_3	−4.73	117	—	—
SO_2CF_3	−4.32	167	—	—
$SOCF_3$	−3.88	143	—	—
SCN	−3.59	127	—	—
SeH	−1.08	169	—	—
$SeCH_3$	−1.31	110	−1.32	—
SeCN	−4.01	124	−3.91	—
$Si(CH_3)_3$	0.44	0	0	0

* The positive sign is taken for groups X that are positive poles of the dipole
R−X. The angle Θ is defined in the following way: R—$\overset{\theta}{\diagdown}$X, where R is
C_6H_5 or CH_3 and X is a polar group. $\Theta > 90°$ for compounds with negative
group moments and $\Theta < 90°$ for compounds with positive group moments.

determination of the vectors of the group moments of irregular
groups X is carried out on the basis of an analysis of the dipole
moment of the p-substituted compound $X−C_6H_4−Y$, where Y is a
regular substituent, usually CH_3 or F, Cl, or Br [54, 55]. Here the
main problem is to evaluate the angle θ, since the scalar magnitude
of the group moment is the same as the dipole moment of C_6H_5X.
The most accurate values of θ are obtained by the statistical treatment of values of the dipole moments for several derivatives with

TABLE 11. Group Moments (in the Gas Phase)

Group X	$\mu_{C_6H_5-X}$, D	μ_{CH_3-X}, D	Group X	$\mu_{C_6H_5-X}$, D	μ_{CH_3-X}, D
CH_3	0.37	0	I	—1.71	—1.70
CF_3	—2.86	—2.35	OH	1.40	—1.70
CCl_3	—	—1.77	OCH_3	—	—1.29
CN	—4.14 *	—3.97	NH_2	1.48	—1.28
CHO	—	—2.72	$N(CH_3)_2$	1.61	—0.61 *
$COCH_3$	—3.00	—2.90 *	NO_2	—4.19	—3.50
F	—1.61	—1.85	SH	—	—1.26 *
Cl	—1.76	—1.86	SCH_3	—	—1.50 *
Br	—1.64	—1.82			

* From the microwave spectrum.

different Y's [56]. If X is an electron-donating group, data for compounds in which Y is also an electron–donating group must be used in the calculations, and conversely. The basis of the calculations is taken from formulas (III.5) and III.6).

Lutskii [57, 58] recommends the determination of the angle θ from the values of the dipole moments of the p–disubstituted derivatives $X-C_6H_4-X$ by calculation from formula (III.12) on the assumption of free rotation round the $C-X$ bonds. Although such rotation does not take place for aromatic compounds, the values of θ taken from (III.12) agree closely with the results of other calculations.*

As has been shown previously (Chapter II), dipole moments depend fundamentally on the conditions of determination. Measurements are carried out most frequently in benzene at 25°C. To interpret the results of these measurements the group moments collected in Table 10 must be used.† Table 11 gives values of the group moments from the results of measurements in the gas phase.

* The applicability of formula (III.12), based on the idea of free rotation, for the calculation of θ can be explained by the fact that the dipole moments of p-$C_6H_4X_2$ calculated from (III.12) on the assumption of an equimolar equilibrium between two cis- and trans-conformations, one of which (trans) is nonpolar, are identical.

† Table 10 gives the results of the most recent or most accurate determinations of the dipole moments. The magnitudes of the angles θ for irregular groups are averaged from several determinations or are given on the basis of the most accurate data [47, 56-67].

TABLE 12. Dipole Moments of Disubstituted Benzenes

Compound	Polar groups	$\theta°$	Angle between the group moments	μ, D	
				calcu-lated	experi-mental
Dichlorobenzene					
o-	Cl, Cl	0	60	2.76	2.27
m-		0	120	1.59	1.48
p-		0	180	0	0
Dinitrobenzene					
o-	NO$_2$, NO$_2$	0	60	6.93	6.00
m-		0	120	4.01	3.88
p-		0	180	0	0
Chloronitroben-zene					
o-	Cl, NO$_2$	0	60	5.00	4.35
m-		0	120	3.49	3.44
p-		0	180	2.42	2.57
Phenylenediamine					
o-	NH$_2$, NH$_2$	48	60	2.38	1.48
m-		48	120	1.90	1.80
p-		48	180	1.60	1.60
Dimethoxyben-zene					
o-	OCH$_3$, OCH$_3$	72	60	1.86	1.38
m-		72	120	1.63	1.76
p-		72	180	1.72	1.70
Anisidine					
o-	OCH$_3$, NH$_2$	72; 48	60	2.10	1.45
p-		72; 48	180	1.78	1.80
Chloroaniline					
o-	Cl, NH$_2$	0; 48	120	1.80	1.77
m-		0; 48	60	2.55	2.66
p-		0; 48	0	2.85	2.92

2. Vectorial Additive Method for Calculating Dipole Moments of Molecules from Bond and Group Moments

Calculations of dipole moments of various structures are based on the rules of vector algebra. It is desirable to distinguish two main structural types of compounds and to give the relation (III.1) applicable to them in a form suitable for direct calculations.

Molecules Containing Two Rigidly Fixed Polar Groups

In this case, the number of terms in the sum of (III.1) is limited to two, and formula (III.1) reduces to the combination of two vectors directed relatively to one another at an angle φ:

$$\mu = \left(\mu_1^2 + \mu_2^2 + 2\mu_1\mu_2 \cos \varphi\right)^{\frac{1}{2}} \tag{III.5}$$

Knowing the group moments μ_1 and μ_2 and having established the angle φ from the geometry of the molecule, it is easy to calculate the dipole moment μ. Examples of such calculations for o-, m-, and p-substituted benzenes using the group moments from Table 10 are given in Table 12.

If the angle φ cannot be determined on the basis of simple considerations as, for example, in the case of disubstituted benzenes, the more general methods of calculation given below must be used.

Molecules Containing any Number of

Rigidly Fixed Polar Groups

Calculations of the dipole moments of molecules containing several polar groups may, of course, be carried out by combining all the group moments in pairs and then proceeding in the same way with the resulting vectors. The method of calculation by projecting the vector of a dipole moment on arbitrarily chosen coordinate axes is considerably more convenient:

$$\mu = \left(m_x^2 + m_y^2 + m_z^2\right)^{\frac{1}{2}} \qquad \text{(III.6)}$$

The projection of a vector of a dipole moment on the appropriate axis is composed of the sum of the projections on this axis of the vectors of all the group moments. Thus, for a molecule including n polar groups

$$\mu = \left[\left(\sum_{i=1}^{n} m_{xi}\right)^2 + \left(\sum_{i=1}^{n} m_{yi}\right)^2 + \left(\sum_{i=1}^{n} m_{zi}\right)^2\right]^{\frac{1}{2}} \qquad \text{(III.7)}$$

In order to calculate from formula (III.7), we must know the angles at which the vectors of the group moments are inclined to the axes of coordinates. In some cases, particularly for planar molecules, these angles are easy to calculate when the valence angles are known. As an example let us consider the calculation of the dipole moment of 4,5-dichloro-1,3-bis(trifluoromethyl)-benzene:

CF$_3$ (1) $m_x = 2.54$; $m_y = 0$

CF$_3$ (3) $m_x = -2.54 \cos 60°$; $m_y = 2.54 \sin 60°$

Cl (4) $m_x = -1.59$; $m_y = 0$

Cl (5) $m_x = -1.59 \cos 60°$; $m_y = -1.59 \sin 60°$

$$\sum m_{xi} = -1.12\ D \qquad \sum m_{yi} = 0.83\ D$$

$$\mu_{\text{calc}} = \sqrt{(-1.12)^2 + 0.83^2} = 1.40\ D$$

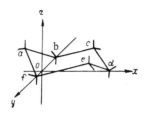

Fig. 16. Scheme for the calculation of the atomic coordinates of the cyclohexane
molecule.

The experimental values
from the data of various authors
[48] are 1.51–1.62 D.

The scheme of calculating
the dipole moment of markedly
nonplanar molecules, for example
alicyclic compounds or hindered
configurations of aliphatic compounds is somewhat more complicated. In this case two methods
for finding the angles of inclination of the vector of the moments
of the polar groups to the axes of coordinates may be proposed.

The first is based on a consideration of the geometry of the
molecular structure constructed for point atomic models, the
Dreiding model [68] being the best. The model of the molecule is
oriented relative to the coordinate axes selected and is projected
on a sheet of millimeter paper arranged in the plane of the coordinates, which gives the atomic coordinates directly. However, this
method is associated with fairly sensitive errors of measurement.
Moreover, it does not enable the calculations to be carried out for
strained structures in which the angles between the bonds differ
appreciably from the tetrahedral and trigonal angles.

In view of this, for accurate calculations an analytical method
of determining the direction cosines of the angles of inclination of
the polar groups to the axes of coordinates must be used [69–72].
The calculation is made up of three stages: 1) the accurate assignment of the geometry and the arrangement of the molecule in the
coordinate axes selected; 2) calculation of the coordinates of the
atoms forming the skeleton of the molecule; 3) calculation of the
direction cosine of the bonds formed by the skeletal atoms with the
substituents.

We may illustrate this scheme by the calculation of the atomic coordinates of the cyclohexane system [69].

1) We place the molecule of cyclohexane in a system of coordinates as shown in Fig. 16. We may regard all the angles
as tetrahedral and the lengths of all the carbon–carbon bonds
as 1.55 Å.

2) The coordinates (x, y, z) of the atoms b and f are evidently: b(-1.266; 0; 0) and f($+1.266$; 0; 0), since

$$\upsilon_i = ob = 1.55 \sin \frac{109°28'}{2} = 1.266$$

We may write the coordinates of the other atoms a(0; $-y_1$; z_1); c(-1.266; y_2; z_1); d(0; y_1+y_2; 0); e(1.266; y_2; z_1) and determine the vectors.

$$\vec{ba} = 1.266\vec{i} - y_1\vec{j} + z_1\vec{k}$$

$$\vec{bc} = y_2\vec{j} + z_1\vec{k}$$

$$\vec{af} = 1.266\vec{i} + y_1\vec{j} - z_1\vec{k}$$

Now we make use of the formula for the scalar product of two vectors expressed in terms of their projections on the axes of coordinates:

$$\vec{a} \cdot \vec{b} = |\vec{a}| \cdot |\vec{b}| \cos \varphi = x_a x_b + y_a y_b + z_a z_b \qquad \text{(III.8)}$$

$$\vec{ba} \cdot \vec{bc} = -y_1 y_2 + z_1^2 = 1.55^2 \cos 109°28' \qquad \text{(III.8a)}$$

$$\vec{ab} \cdot \vec{af} = -1.266^2 + y_1^2 + z_1^2 = 1.55^2 \cos 109°28' \qquad \text{(III.8b)}$$

$$\vec{bc} \cdot \vec{bc} = y_2^2 + z_1^2 \qquad \text{(III.8c)}$$

whence $y_1 = 0.729$, $y_2 = 1.46$, and $z_1 = 0.521$.

3) Having determined in this way the coordinates of all the skeletal atoms, we can find the direction cosines of the angles of inclination of any bond to the axes of coordinates from the well-known formula for determining the angle between two vectors:

$$\cos \varphi = \frac{\vec{a} \cdot \vec{b}}{|\vec{a}| \cdot |\vec{b}|} = \frac{a_x b_x + a_y b_y + a_z b_z}{[(a_x^2 + a_y^2 + a_z^2)(b_x^2 + b_y^2 + b_z^2)]^{\frac{1}{2}}} \qquad \text{(III.9)}$$

Here as the vector $\vec{a}$ we may take any vector, for example $\vec{af}$ (see Fig. 16) and as $\vec{b}$ an individual vector $\vec{i}, \vec{j},$ or $\vec{k}$, depending on the axis of coordinates with respect to which the angle of slope is defined.

The values of the direction cosines of the bonds obtained are used to calculate the projections of the moments of these bonds on the coordinate axes. In the case considered, the moment of the C$-$C bond is zero, and the values of the direction cosines of the C$-$C bonds are required only to calculate the direction cosines of

TABLE 13. Coordinates of the Skeletal Atoms and Unit Vectors of the Substituents of Cyclohexane according to Eliel et al. [73]

Coordinates of the atoms in Å				Unit vectors of the substituents			
atom	x	y	z	substituent	x	y	z
1	−1.325	0	−0.651	1a	0.223	0	−0.975
				1e	−0.997	0	0.077
2	−0.764	−1.263	0	2a	−0.333	−0.067	0.940
				2e	−0.333	−0.804	−0.492
3	0.764	−1.263	0	3a	0.333	−0.067	−0.940
				3e	0.333	−0.804	0.492
4	1.325	0	0.651	4a	−0.223	0	0.975
				4e	0.997	0	−0.077
5	0.764	1.263	0	5a	0.333	0.067	−0.940
				5e	0.333	0.804	0.492
6	−0.764	1.263	0	6a	−0.333	0.067	0.940
				6e	−0.333	0.804	−0.492

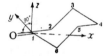

TABLE 14. Coordinates of the Skeletal Atoms and Unit Vectors of the Substituents of Cyclohexanone according to Eliel et al. [73]

Coordinates of the atoms in Å				Unit vectors of the substituents			
atom	x	y	z	substituent	x	y	z
1	0	0	0	=O	−1	0	0
2	0.795	1.272	0	2a	0.433	0.123	−0.893
				2e	−0.614	0.777	0.141
3	1.868	1.265	1.090	3a	−0.436	0.063	0.898
				3e	0.585	0.804	−0.103
4	2.726	0	1.028	4a	0.533	0	−0.846
				4e	0.649	0	0.761
5	1.868	−1.265	1.090	5a	−0.436	−0.063	0.898
				5e	0.585	−0.804	−0.103
6	0.795	−1.272	0	6a	0.433	−0.123	−0.893
				6e	−0.614	−0.777	0.141

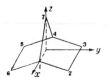

TABLE 15. Coordinates of the Skeletal Atoms and Unit Vectors of the Substituents of Norbornane (bicyclo[2,2,1]heptane) according to Wilcox [70]

Coordinates of the atoms in Å				Unit vectors of the substituents			
atom	x	y	z	substituent	x	y	z
1	1.143	0	0	1	0.927	0	0.376
2	0.770	1.217	−0.868	2-exo	0.349	0.842	0.411
				2-endo	0.349	−0.115	−0.930
3	−0.770	1.217	−0.868	3-exo	−0.349	0.842	0.411
				3-endo	−0.349	−0.115	−0.930
4	−1.143	0	0	4	−0.927	0	0.376
5	−0.770	−1.217	−0.868	5-exo	−0.349	−0.842	0.411
				5-endo	−0.349	0.115	−0.930
6	0.770	−1.217	−0.868	6-exo	0.349	−0.842	0.411
				6-endo	0.349	0.115	−0.930
7	0	0	1.032	7 (right)	0	0.824	0.566
				7 (left)	0	−0.824	0.566

the bonds formed by the skeletal atoms with their substituents. For this purpose, unit vectors (vectors, the scalar magnitudes of which are equal to unity and the projections of which are, therefore, equal to the direction cosines) are isolated that are directed along the line of the bond to the substituent, and the scalar products of the individual vectors of the substituents and the individual vectors of the skeletal bonds are written in accordance with (III.8). The solution of these three equations of type (III.8a–c) gives the required direction cosines of the unit vectors of the substituent.

Such calculations have been carried out for many alicyclic systems. In many cases they were based on the point geometry of the molecules determined from electron-diffraction or X-ray measurements.

Tables 13–15, compiled from literature data [70–73], give the atomic coordinates and unit vectors of the substituents for molecules based on cyclohexane and norbornane. The values for cyclohexane and cyclohexanone have been refined somewhat as compared with the data based on the calculation discussed above, and the results of later structural studies of cyclohexane derivatives have been taken into account. The data for norbornane were obtained on the basis of the length of the carbon—carbon bond of 1.54 Å; the valence angles were calculated by minimizing the strain energy of the bicyclo[2,2,1]heptane system.

We may make use of the data of Table 13 to calculate the dipole moments of the 1,2-diaxial (I) and 1,2-diequatorial (II) isomers of dichlorocyclohexane.

Since the molecule of cyclohexane is nonpolar, the dipole moment of compounds (I) and (II) can be regarded as the sum of the vectors of the moments of the two C—Cl bonds with the moments of the C—H bonds deducted. Each of these moments can be taken as equal to the moment of chlorocyclohexane (2.24 D). Then, for the diaxial isomer the projections of the vector of the dipole moment on the x, y, z axes in the orientation corresponding to the scheme of Table 13 are (in D):

$$\sum_i m_{xi} = 2.24 \cdot 0.223 + 2.24 \cdot (-0.333) = -0.25$$

$$\sum_i m_{yi} = 2.24 \cdot 0 + 2.24 \cdot (-0.067) = -0.15$$

$$\sum_i m_{zi} = 2.24 \cdot (-0.975) + 2.24 \cdot 0.940 = -0.08$$

Hence, according to (III.7), $\mu_{calc} = 0.32$ D.

For the diequatorial isomer:

$$\sum_i m_{xi} = 2.24 \cdot (-0.997) + 2.24 \cdot (-0.333) = -2.98$$

$$\sum_i m_{yi} = 2.24 \cdot 0 + 2.24 \cdot (-0.804) = -1.80$$

$$\sum_i m_{zi} = 2.24 \cdot 0.077 + 2.24 \cdot (-0.492) = -0.93$$

and $\mu_{calc} = 3.61$ D. The experimental value of the dipole moment of trans-1,2-dichlorocyclohexane of 2.63 D (benzene, 25°C) is intermediate between the two values calculated. This is in agreement with the assumption of the conformational lability of the cyclohexane ring, leading to the equilibrium I ⇌ II.

Molecules Containing Two Freely Rotatable Polar Groups

If the molecule contains two polar groups with the moments μ_1 and μ_2 that rotate freely about axes a_1 and a_2, respectively, the angles formed by the directions of the axes of rotation and the vectors of the group moments being θ_1 and θ_2 and these axes forming with one another the angle ω_{12}, the overall moment is calculated from the formula [74, 75]:

$$\mu = \left(\mu_1^2 + \mu_2^2 + 2\mu_1\mu_2 \cos\theta_1 \cos\theta_2 \cos\omega_{12}\right)^{\frac{1}{2}} \qquad (\text{III}.10)$$

As an example, let us consider the value of the dipole moment of m-aminophenol on the assumption of the free rotation of the amino and hydroxy groups about the Ar−N and Ar−O bonds, respectively. The magnitudes of μ_1, μ_2, θ_1, and θ_2 are taken from the data of Table 10.

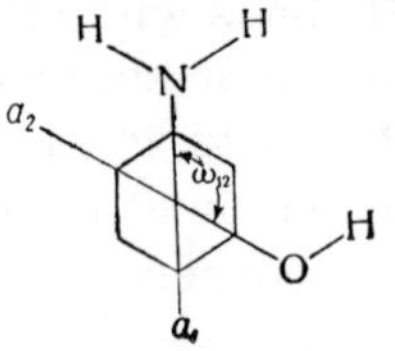

μ_1 (NH$_2$) = 1.53 D θ_1 = 48.5°
μ_2 (OH) = 1.55 D θ_2 = 90°
ω_{12} = 120°

By substituting these values in (III.10), we obtain $\mu_{\text{calc}} =$ 2.18 D. The experimental value of the dipole moment of m−aminophenol is 2.19 D (in dioxane) [58]. This agreement still cannot be regarded as a proof of the existence of free rotation of the groups, since the same dipole moment can be calculated on the assumption of equally probable configurations of the NH$_2$ and OH groups formed by rotation through 180° about the axes a_1 and a_2 (four equally probable molecular conformations). It will be shown below (Chapter IV) that free rotation is absent not only in groups conjugated with unsaturated systems but also in aliphatic derivatives. Thus, the agreement of the dipole moment calculated according to (III.10) with the experimental value must be regarded as an indication of equal populations of molecular conformations differing from one another by the angle of rotation of the polar group (180°).

Table 12 gives some examples of the calculation of the dipole moments of disubstituted benzenes with irregular polar groups according to formula (III.10). It can be seen that for m- and p−disubstituted benzenes the agreement of the calculated and experimental values is extremely good. The cause of the nonagreement of the calculated and experimental values for the o-derivatives will be discussed in Chapter V.

Formula (III.10) can be simplified for use with certain special cases. For example, for the case of one rigidly fixed ($\theta_1 = 0°$) and one freely rotating substituent

$$\mu = \left(\mu_1^2 + \mu_2^2 + 2\mu_1\mu_2 \cos \theta_2 \cos \omega_{12}\right)^{\frac{1}{2}} \qquad \text{(III.11)}$$

The simplest case is the free rotation of two similar polar groups relative to a single axis: $\mu_1 = \mu_2$, $\theta_1 = \theta_2 = \theta$, $\omega_{12} = 180°$. Then, (III.10) reduces to the Williams equation (III.12) [16]:

$$\mu = \sqrt{2}\,\mu_1 \sin \theta \qquad \text{(III.12)}$$

The dipole moments of terephthalaldehyde (I) and dimethyl terephthalate (II) calculated, for example, from the data of Table 10 are respectively, 2.34 and 2.43 D and agree excellently with the experimental values [48]:

OHC—⟨ ⟩—CHO   H₃COOC—⟨ ⟩—COOCH₃

I   II

$\mu = 2.35\ D$ $\mu = 2.40\ D$

Molecules Containing Any Number of Freely Rotatable Polar Groups and Axes of Rotation

If the molecule contains several polar groups, the relative position of which changes continuously with very high frequency because of free rotation about several bonds, the dipole moment determined experimentally is the value averaged with respect to all the configurations of the molecule arising as the result of the free rotation.

Eyring [77] has shown that in this case the average value of the square of the resulting dipole moment of the molecule $\overline{\mu^2}$ can be represented in the form of the sum of the squares of the moments of all n polar groups or bonds with moments μ_i and all the doubled products in pairs of these moments multiplied by the products of the cosines of all the angles θ_k by which the bonds connecting a given pair of dipoles change direction on passing along the chain from one dipole of a given pair to the other:

$$\overline{\mu^2} = \sum_{i=1}^{n} \mu_i^2 + 2 \sum_{i=1}^{n} \sum_{s < i} \prod_{k=i}^{s+1} \cos \theta_k \mu_i \mu_s \qquad \text{(III.13)}$$

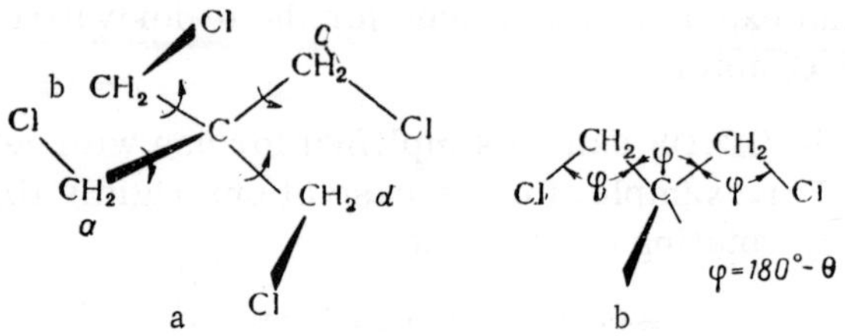

Fig. 17. Scheme for calculating the dipole moment of
tetra(chloromethyl)methane.

The calculation can be illustrated for the case of the dipole moments of tetra(chloromethyl)methane and α-bromo-ω-cyanodecane (Fig. 17).

The molecule of tetra(chloromethyl)methane contains four similar polar substituents the group moment of which $\mu_1 = 1.87$ D. If it is assumed that all the valence angles of the carbon atom are tetrahedral, i.e. 109.5°, the angles θ (Fig. 17) are 180–109.5°. Let us now take into account the fact that when two dipoles in a common chain have opposite directions, if the chain is drawn out as in Fig. 17b, one of them must have a minus sign when the products in pairs are summed.

It may be considered that the molecule of tetra(chloromethyl)methane is composed of six chains, as shown in Fig. 17a, in the combinations: ab, ac, ad, bc, bd, cd. Bearing this in mind, it is not difficult, by using equation (III.13) to obtain the following expression for the dipole moment of tetra(chloromethyl)methane:

$$\bar{\mu}^2 = 4\mu_1^2 + 6\left(-2\mu_1^2 \cos^3 70.5°\right)$$

from which $\mu_{calc} = 3.52$ D. This magnitude differs considerably from the experimental value of 0.63 D (in benzene) [78], which makes the assumption of the free rotation of all the bonds in tetra-(chloromethyl)methane unrealistic.

Conversely, the dipole moment of α-bromo-ω-cyanodecane calculated from formula (III.13) agrees well with the experimental value of 3.95 D (benzene, 75°C) [79].

$$\begin{array}{cccccc}
CH_2 & CH_2 & CH_2 & CH_2 & CH_2 & CN \\
\diagup \diagdown & \diagup \diagdown & \diagup \diagdown & \diagup \diagdown & \diagup \diagdown & \diagup \\
Br & CH_2 & CH_2 & CH_2 & CH_2 & CH_2
\end{array}$$

$$\bar{\mu}^2 = \mu_{Br}^2 + \mu_{CN}^2 - 2\mu_{Br}\mu_{CN}\cos^{10} 70.5°$$

$$\mu = 3.92D$$

Eyring's formula is suitable for calculating dipole moments of molecules having a chain structure with fixed angles between the links and free rotation of each successive link about the axis formed by its bond with the preceding link. A more general formula is that (III.14) proposed by Zahn [80]. Apart from the cases of chain-like molecules discussed above, formula (III.14) can be used to calculate dipole moments of molecules with the structure of a rigid nucleus to which freely rotating irregular polar groups are attached, for example derivatives of benzene, naphthalene, and aromatic heterocycles.

$$\bar{\mu}^2 = \left(\sum_{i=1}^{n}\mu_{pi}\right)^2 + \left(\sum_{i=1}^{n}\mu_{ki}\right)^2 \tag{III.14}$$

Here, μ_{pi} and μ_{ki} are the components of the i-th group moment directed parallel and perpendicular to the axis of rotation and n is the number of freely rotating polar groups.

For a particular case (trisubstituted benzenes possessing freely rotating polar groups with group moments of μ_1, μ_2, and μ_3, with the vectors of the group moments making angles of θ_1, θ_2, and θ_3 with the axes of rotation and lying in the plane of the benzene nucleus and the axes of rotation forming the angles ω_{12} and ω_{13} with one another) the following relation is valid [74]:

$$\mu^2 = \mu_1^2 + \mu_2^2 + \mu_3^2 + 2\mu_1\mu_2\cos\theta_1\cos\theta_2\cos\omega_{12}$$

$$+ 2\mu_2\mu_3\cos\theta_2\cos\theta_3\cos(\omega_{12}+\omega_{13}) + 2\mu_1\mu_3\cos\theta_1\cos\theta_3\cos\omega_{13} \tag{III.15}$$

By using formula (III.15), let us calculate the dipole moment of 1,3,5-triaminobenzene on the assumption of free rotation of the amino groups which is equivalent to the assumption of the equivalence and equiprobability of all the possible conformations of the molecule:

$$\mu_1 = \mu_2 = \mu_3 = 1.53\,D$$
$$\theta_1 = \theta_2 = \theta_3 = 48.5°$$
$$\omega_{12} = \omega_{13} = 120°$$

$$\mu^2 = 3\,(1.53)^2 + 3\,[2\,(1.53)^2\,\cos^2 48.5°\cos 120°]$$
$$\mu = 1.99\,D$$

Molecules Containing Any Number of Rigidly
Fixed and Freely Rotatable Polar Groups

This most general case has recently been considered by Gil-man [81]. For molecules with a rigid skeleton of bonds to which one or more regular or irregular polar groups freely rotating about their bonds with the rigid nucleus are attached, he derived the equation:

$$\bar{\mu}^2 = \mu_0^2 + \sum_{i=1}^{n} \mu_i^2 + 2 \sum_{i=1}^{n} \mu_i \left(\mu_{0x}a_{ix} + \mu_{0y}a_{iy} + \mu_{0z}a_{iz}\right)\cos\theta_i$$

$$+ 2 \sum_{i \neq j}^{n} \mu_i \mu_j \left(a_{ix}a_{jx} + a_{iy}a_{jy} + a_{iz}a_{jz}\right)\cos\theta_i \cos\theta_j \qquad (III.16)$$

Here μ_0 is the resultant dipole moment of all the fixed polar groups calculated in the manner shown above; μ_{0x}, μ_{0y}, μ_{0z} are the projections of $\vec{\mu}_0$ on the coordinate axis selected; μ_i and μ_j are the group moments of the freely rotating polar groups the total number of which is n; θ_i and θ_j are the angles formed by the vector of the group moments with the axes of rotation a_i and a_j; and, finally, a_{ix}, a_{iy}, and a_{iz} are the unit vectors of the i-axis of rotation in the system of coordinates selected.

The use of relation (III.16) can be illustrated by calculating the dipole moments of the exo and endo isomers of norborneol (I) and (II)

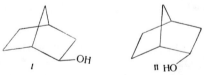

Since, in this case, there is only one freely rotating polar group (OH), equation (III.16) is substantially simplified:

$$\overline{\mu}^2 = \mu_0^2 + \mu_{OH}^2 + 2\mu_{OH}(a_x\mu_{0x} + a_y\mu_{0y} + a_z\mu_{0z})\cos\theta \tag{III.17}$$

The dipole moment of the rigidly fixed part of the molecule can be taken as equal to the dipole moment of norbornane, which, according to various authors, is from 0.15 D [82] to 0.24 D [83]. We shall use the average value 0.2 D. In view of the symmetry of norbornane it is obvious that the components of μ_0 will be $\mu_{0x} = \mu_{0y} = 0$ and $\mu_{0z} = \pm 0.2$ D. The moment of the OH group and the angle θ will be taken, following Tanner and Gilman [84], as 1.60 D and

63°. The components of the unit vector $\vec{a}$ have been taken from Table 15. For $m_z = +0.20$ D and the exo isomer we obtain

$$\overline{\mu}^2 = 0.2^2 + 1.6^2 + 2 \cdot 1.6 \,(0.349 \cdot 0 + 0.842 \cdot 0 + 0.411 \cdot 0.2)\cos 63°$$
$$\mu = 1.65 \ D$$

For the endo isomer

$$\overline{\mu}^2 = 0.2^2 + 1.6^2 + 2 \cdot 1.6 \,(0.349 \cdot 0 - 0.115 \cdot 0 - 0.930 \cdot 0.2)\cos 63°$$
$$\mu = 1.53 \ D$$

The experimental values for the exo and endo isomers are 1.63 and 1.66 D [83]. Some other examples of calculations by formulas (III.16) and (III.17) have been given by Gilman [81].

Molecules with Hindered Rotation of the Polar Groups

The rotation of individual fragments of a molecule relative to the bonds joining them as axes of rotation is unopposed only if the kinetic energy of the molecule exceeds the potential barrier to rotation. This is usually the case only at fairly high temperatures. In all other cases, the rotation is hindered and degenerates into torsional oscillations of the polar groups relative to some energetically most favorable configuration: the angle of rotation $\varphi < 2\pi$ (Fig. 18), which corresponds to the existence of a number of rotational isomers the conformations of which are limited by the value of the angle φ.

In calculating the effective dipole moment of such a molecule we must know the potential function of the internal rotation $V(\varphi)$. If this function is known, then according to the Boltzmann law, the unnormalized probability density for the molecule to have the dipole moment μ is known:

$$P(\mu, \varphi) = e^{\frac{-V(\varphi)}{kT}} \tag{III.18}$$

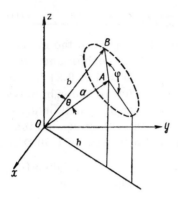

Fig. 18. Coordinate system of a rotating vector.

Then the mean square moment of the molecule at a given temperature T (°K) is determined by means of the equation

$$\bar{\mu}^2 = \frac{\displaystyle\int_{\varphi_1}^{\varphi_2} \mu^2 e^{-\frac{V(\varphi)}{kT}} d\varphi}{\displaystyle\int_{\varphi_1}^{\varphi_2} e^{-\frac{V(\varphi)}{kT}} d(\varphi)} \qquad (III.19)$$

where φ_1 and φ_2 correspond to the limiting values of the angle of rotation; for example, in rotation relative to a single bond, as in the 1,2-dihaloethanes, $\varphi_1 = 0$ and $\varphi_2 = 2\pi$.

Examples of the practical application of equation (III.19) can be found in Chapter IV, Section 5.

In some cases the necessity may arise for calculating a dipole moment for any of the possible orientations of the rotating polar group defined by a definite value of the angle φ. For such a calculation we must know the components of the unit vector of the group moment of the rotating group. For a rotating irregular group with an angle of inclination of the vector of the group moment to the axis of rotation of θ, the components of the unit vector in the hindered configuration are expressed in the following way [81]:

$$b'_x = a_x \cos\theta + [a_x a_z \sin\theta \cos\varphi - a_y \sin\theta \sin\varphi]\frac{1}{h}$$

$$b'_y = a_y \cos\theta + [a_y a_z \sin\theta \cos\varphi + a_x \sin\theta \sin\varphi]\frac{1}{h} \qquad (III.20)$$

$$b'_z = a_z \cos\theta - h \sin\theta \cos\varphi$$

The symbols have the same meanings as in equation (III.16) and are explained in Fig. 18.

Molecules Existing as an Equilibrium Mixture of Several Forms

The molecules of some compounds exist in solutions or in the gas phase as an equilibrium mixture of two or more molecular forms: tautomers, conformers, monomers and dimers, etc. The experimentally determined dipole moment in this case is an average value connected with the dipole moments of the individual forms by the relation

$$\mu^2 = \sum_{i=1}^{n} N_i \mu_i^2 \qquad (III.21)$$

where N_i and μ_i are the relative concentration and the dipole moment of the tautomer, conformer, or the like.

The additivity of the squares of the dipole moments (III.21) follows from the additivity of polarizations (Chapter II).

Calculation from formula (III.21) can be illustrated by the calculation of the effective dipole moment of 2,5-dimethoxynitrobenzene based on the assumption that this molecule exists in solution in the form of a mixture of the four possible equiprobable planar conformations (I)-(IV). It is considered that this assumption is certainly not accurate, since it is clear that these conformations (II) and (IV) will have a greater weight in the composition of the mixture of conformers than (I) and (III) in which the methoxy groups are subject to considerable steric hindrance. Moreover, it will be necessary to take into account the possible presence in the solution of nonplanar conformations of the molecule of 2,5-dimethoxynitrobenzene. However, it is obvious that without special additional information it is impossible to take all these circumstances into account.

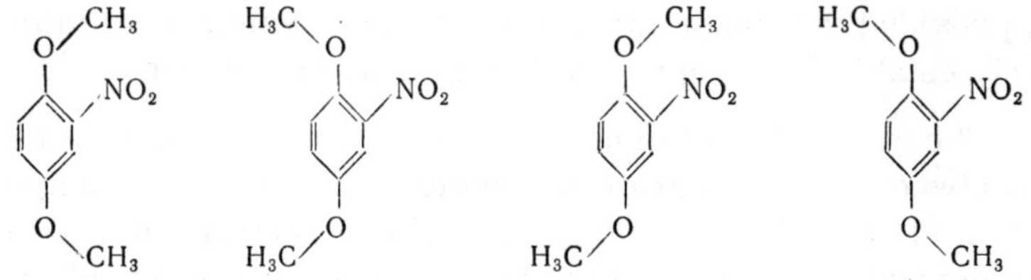

I
$\mu_{calc} = 2.27D$ II
$\mu_{calc} = 6.20D$ III
$\mu_{calc} = 4.01D$ IV
$\mu_{calc} = 4.01\ D$

By using the values of the group moment from Table 10 and calculating the theoretical dipole moments of each conformer, we obtain:

$$\mu^2 = \frac{1}{4}(2.27)^2 + \frac{1}{4}(6.20)^2 + \frac{1}{4}(4.01)^2 + \frac{1}{4}(4.01)^2$$

$$\mu = 4.34\ D$$

The value obtained is close to the experimental value of 4.56 D (benzene, 18°C) [48].

Relation (III.21) is frequently used to solve the converse problem: to calculate the amount of the individual forms in which a molecule exists in solutions. This calculation starts from the experimentally measured effective dipole moment and the calculated moments of the individual forms. Thus, by using the values of the dipole moments of the diaxial and diequatorial isomers of 1,2-dichlorocyclohexane calculated above, it is possible to determine the content of each conformer in benzene at 25°C.

$$2.63^2 = x \cdot 3.61^2 + (1 - x) 0.32^2$$

It follows from this that the content of the diequatorial conformer is 48%.

Graphical Method of Analyzing the Data of the

Vectorial Additive Scheme

Exner and Jehlicka [85] have proposed a useful graphical method of analyzing data based on the vectorial scheme which has been used [86, 87] to calculate the dipole moment of a number of aromatic compounds capable of existing in solutions in the form of one out of several theoretically possible conformations or mixtures of them. The value of this method is the possibility of evaluating the degree of suitability of the bond and group moments used in the calculations of the dipole moment of a given compound and, provided that these moments have been selected suitably, of clearly establishing a preferable molecular structure.

The principle of the method consists in plotting the relationship between the squares of the dipole moments of the compound under study calculated for each possible conformation μ_H^2 and those of its derivative μ_X^2 formed by the replacement (preferably in the p-position) of one of the hydrogen atoms in the phenyl nucleus by a regular polar substituent X (NO_2, Br, F, . . .). Here, each conformation corresponds to a single point on the graph. The experimental values of μ_H^2 and μ_X^2 form a separate point which must coincide approximately with one of the points on the graph and this enables the configuration of the molecule to be determined.* If this

---

* A necessary assumption here is that the substituent X located remote from the functional group attached to the aryl ring has no specific influence on its configuration, i.e., the configuration of the molecule substituted in the phenyl nucleus and of the unsubstituted molecule are analogous.

is not the case, there are two possible reasons for a discrepancy between the experimental and calculated figures: 1) incorrect selection of the bond and group moments or the valence angles, and, 2) the conformations considered do not correspond to the actual structure of the molecule in solution.

If systematic changes in the group and bond moments of the valence angles do not enable the calculated value to be brought near to the experimental figure, only the latter possibility remains, which generally corresponds to one of the following causes: 1) the compound exists in solution as an equilibrium mixture of two conformations, or 2) there is free rotation round one of the bonds. Under these conditions, the relative position of the experimental point on the graph with respect to the points corresponding to the limiting conformations enables us to determine the composition of the equilibrium mixture or some geometrical parameters of the compound under study.

Let us consider two examples of the application of the graphical method of analysis of the results of vectorial calculation. The first of them is the determination of the conformation of phenylacetic acid (I).

$$C_6H_5CH_2COOH \qquad\qquad C_6H_5NH-\underset{\underset{O}{\|}}{C}-CH_3$$

I   II

By resolving into the bond moments the dipole moment of aliphatic carboxylic acids (about 1.7 D [48]), we can calculate the moments for any conformations of (I) formed by the rotation of the carboxyl round the C_1-C_2 bond (Fig. 19). Points A and B in the figure correspond to the limiting plane conformations. The experimental point E falls on the line AB in the region of an angle of rotation of 60° corresponding to the stable gauche conformation. The position of the experimental point may also correspond to the equilibrium $A \rightleftarrows B$ displaced in the B direction in the ratio $\overline{BE}/\overline{EA}$. The choice between these two possibilities can be made only with the aid of additional information. In this case, from steric considerations, in agreement with the results of other methods of investigation, the idea of the presence of the gauche conformation is the correct one.

When a hydrogen atom rotates about a $C-O$ bond, a number of new conformations arise in the limit conformations C and D

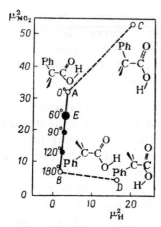

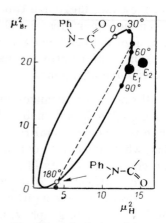

Fig. 19. Calculation of the dipole moments of various conformations of phenylacetic and
p-nitrophenylacetic acids [85]:
E is the point corresponding to
the experimental results.

Fig. 20. Calculation of the dipole moments of various conformations of acetanilide and
p-bromoacetanilide [85].

having the s–cis structure (Chapter IV, Section 3). The broken
lines in Fig. 19 show the dependence of $\mu_{NO_2}^2 - \mu_H^2$ on the angle of
rotation round the C—O bond. It can be seen that the experimental
point is remote from these lines, from which we can deduce the
unsuitability of the s–cis configuration of the carboxyl.

Figure 20 shows a somewhat more complex case. The el-
lipse drawn in the figure represents $\mu_{Br}^2 - \mu_H^2$ as a function of the
angle of rotation around the C—O bond in the molecule of acetanilide (II) for which the pyramidal configuration of the nitrogen atom
is assumed. The positions of the experimental points E_1 (benzene)
and E_2 (dioxane) correspond most closely to a conformation in
which the angle between the C_{Ar}—N—C and N—C $=$ O planes is
about 80°. However, accurate assignment is difficult here because
not even the plane configuration with a trigonal nitrogen can be excluded: the broken line corresponding to the angular dependence
of the magnitude $\mu_{Br}^2 - \mu_H^2$ passes fairly close to E_1 and E_2.

A number of other examples are given by Exner et al. [85-87].
Summarizing, it can be said that the graphical method somewhat
broadens the possibility of the vectorial scheme mainly because of

its graphic and systematic nature, but it is fully based on the principles of this scheme and has the same internal defects, which are
discussed below.

3. Critical Observations on the Vectorial Additive Scheme and Methods for Its Improvement

A consideration of the data given above shows that with certain assumptions concerning the structure of a molecule, the theoretical calculation of its dipole moment does not offer great difficulties. This calculation, however, is necessarily based on the
magnitudes of the bond and group moments, the use of which for
different types of molecules requires great caution.

Apart from some uncertainties in the evaluation of bond and
group moments and the practical impossibility of allowing reliably
for the dipoles of unshared electron pairs, to which attention was
drawn in Section 1 of the present chapter, we may mention above
all that the magnitudes of the bond and group moments given in
Table 17, Table 19, and Table 20 were calculated from the dipole
moments of molecules with only one polar group and, consequently
do not take into account the possibility of the mutual influence of
the atoms and groups in more complex molecules. The nature of
this interaction may differ according to the structure of the molecule: apart from the inductive interaction between the polar
groups which is possible for all types of molecules, in molecules
with π-electrons there are various conjugation effects. These effects of electronic displacement are reflected in the degree of polarization of the bonds, which may lead to appreciable changes in
their dipole moments as compared with the unperturbed magnitudes.

This situation is not, of course, a feature of the additive
method of calculating dipole moments alone but is also characteristic for all other additive schemes and ideas (for example, for
molecular refractions, magnetic susceptibilities, IR and NMR
spectra, etc.). The observed deviations from the additive values
are used to demonstrate and analyze effects not included in the
additive magnitude. The methods available at the present time for
the quantitative evaluation of the additional effects of inductive

mesomeric interaction and the effect of polar conjugation of polarized bonds and groups are considered in Chapter V.

The most convenient and usually most accurate method for including specific interaction effects for a given type of compound is to take them into account in implicit form by the use in the calculations of bond and group moments derived from the dipole moments of some initial compounds of the type concerned. In fact, the division of group moments into moments in the aliphatic and aromatic series (Table 10) may itself be regarded as an example of the practical use of this device. In calculating the moment of 1,2-dichlorocyclohexane (Section 2) we have likewise used as the moment of the $C-Cl$ bond the moment of chlorocyclohexane, which differs somewhat from the moment given in Table 19. Other examples of the use of this approach are given in Chapter IV, Section 5. The desirability of this method is shown, in particular, by the fact that even in saturated systems the relative configuration of the polar groups may lead to differences in their moments sometimes amounting to 1-2 D [88].

At the present time, for several classes of compounds their own systems of bond and group moments have been created, being based on the resolution of the experimental moments of some compounds specially selected within the class concerned and including all the possible types of bond in various combinations. The most systematically worked out are the schemes of calculation for the dipole moments of the alkanes [89-91] and the polyhalocyclohexanes [92], which are based on Tatevskii's classification of bond types [93].

A further development of the vectorial additive scheme is the calculation of the dipole moment of a complex molecule in the form of the sum of the vectors of the moments of its individual fragments. As an example, we may consider the calculation of the dipole moment of o-methoxybenzylideneaniline in four possible planar conformations [94]:

I

$\mu_{calc} = 1.07\ D$

II

$\mu_{calc} = 2.23\ D$

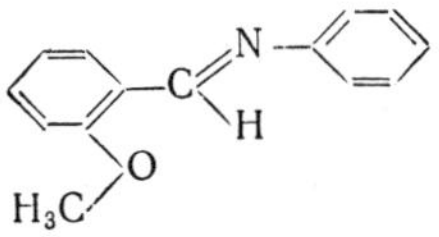

III
$\mu_{calc} = 2.70\ D$

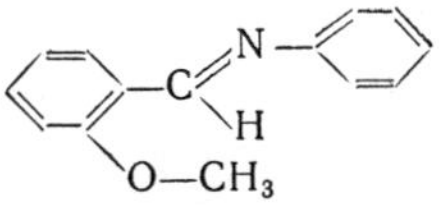

IV
$\mu_{calc} = 1.84\ D$

The moment corresponding to each conformation can be represented as the sum of the vectors of the moments of benzylideneaniline (V) and anisole, the vector of the moment of the latter being oriented relative to the vector of the moment of benzylideneaniline in accordance with the orientation of the anisole nucleus in (I)-(V), i.e., (Ia) for (I), (IIa) for (II), and so on.

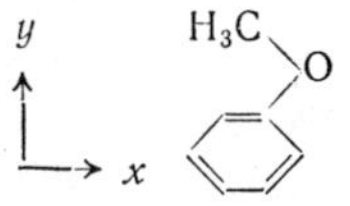

Ia
$m_x = 0.84;$
$m_y = -0.95\ D$

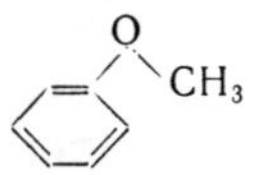

IIa
$m_x = -1\ 24;$
$m_y = 0.25\ D$

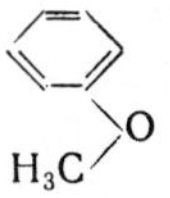

IIIa
$m_x = 0.84;$
$m_y = 0.95\ D$

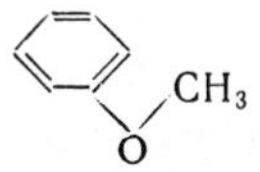

IVa
$m_x = -1\ 24;$
$m_y = -0.25\ D$

The projections of the vector of the moment of anisole on the axes of coordinates are determined in accordance with the data of Table 10 and the projections on x, y of the vector of benzylideneaniline (V) are calculated from the moments of (V), its p-methyl derivative (VI), and the methyl group (0.37 D)

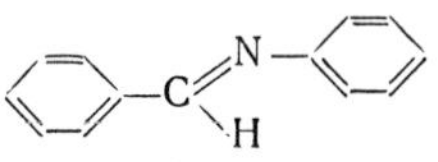

V
$\mu_{exp} = 1.61\ D\ [95]$

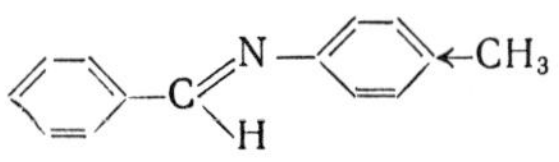

VI
$\mu_{exp} = 1.65\ D\ [96]$

$$x^2 + y^2 = 1.61^2$$
$$(x - 0.37)^2 + y^2 = 1.65^2$$

from which x = 0 and y = 1.61 D.

A combination of the vectors of the moments of anisole (Ia) - (IVa) and benzylideneaniline gives the values of the moments of o-methoxybenzylideneaniline in the conformations (I)-(IV). A comparison of the figures obtained with the experimental figure of 2.87 D [95] is in favor of conformation (III), the predominance of which is due to the fact that it has the lowest steric hindrance.

The advantage of vectorial calculation with respect to fragments is particularly evident in calculations of the dipole moments
of unsaturated heterocyclic compounds. Since the individual bonds
including the heteroatoms form part of the conjugated system of
such compounds, they cannot be assigned definite fixed values of
the bond moments. Similarly, it is illegitimate to calculate these
moments from the dipole moments of aromatic carbocyclic compounds. Even the moments of polar groups linked to a heterocycle
differ substantially from the moments of the same groups in the
aromatic series. For example, the moment of the C−I bond in
2-iodofuran is only 0.56 D [97], while in the aromatic series the
moment of this bond is 1.40 D (Table 10). Similar discrepancies
are characteristic for other heterocyclic compounds (thiophene
[98], pyridine [99], etc.), the values of the group moments depending on their position in the heterocycle. The main cause of such
deviations is, in the general case, the electrical asymmetry of the
heterocycle and the different effective charges on the various carbon atoms in one and the same and in different heterocycles.

As an example of the calculation of the dipole moment of a
heterocyclic compound from the moments of the individual fragments we may give the calculation of the dipole moments of the
(p-nitrophenyl)pyridines (VII)-(IX) [100]. The moment of pyridine,
which is 2.21 D, must be assumed to be directed along a second-
order axis of symmetry in the direction of the unshared electron
pair. The calculated values agree fairly well with the experimental figures.

VII	VIII	IX
$\mu_{calc} = 5.51\ D$	$\mu_{calc} = 3.54\ D$	$\mu_{calc} = 1.78\ D$
$\mu_{exp} = 5.51\ D$	$\mu_{exp} = 3.79\ D$	$\mu_{exp} = 1.93\ D$

In compounds (VII)-(IX), the polar NO_2 group is attached to an aryl
nucleus, which gives grounds for using as its dipole moment the
usual value in the aromatic series (Table 10). Calculation is substantially facilitated also by the fact that the choice of the direction
of the vector of the pyridine moment follows from the symmetry of

the heterocycle. In those cases where there is no such symmetry, the position is greatly complicated. Simplified empirical schemes have been proposed for individual classes of heterocycles which enable the overall dipole moments to be evaluated with an accuracy of 1-1.5 D. A similar scheme has been developed for azines and azoles by Giller and Mazheika [101, 102].

In some cases, where a choice must be made between two or more possible structures on the basis of the experimental dipole moment and where the calculation of the moments expected for these structures is difficult or indeterminate, it may be very useful to make a comparison with the dipole moments of model structures. Thus, the question of the cis or trans configuration of 1-azobis-2-phenylpropane and esters of carboxylic acids is easily solved by comparing their dipole moments with the moments of compounds in which the cis configuration is rigidly fixed (see Chapter IV, Section 3). Another characteristic example of the creation of a suitable model structure is the fixing of a conformationally labile system by introducing the voluminous nonpolar tert-butyl group, which always occupies the spatially more favorable equatorial position and hinders the conversion of the ring (see Chapter IV, Section 5).

4. Quantum-Mechanical Calculations

of the Dipole Moments of Complex Molecules

The second basic possibility for the theoretical calculation of dipole moments, after the vectorial scheme, is a calculation based on the methods of quantum mechanics. In principle, each of the possible or expected structures or conformations for a given compound is characterized by its inherent wave function Ψ and its inherent electronic configuration, the dipole moment of which is defined by the relation

$$\vec{\mu} = -e \int \sum_i \vec{r}_i \, [\Psi \, (1, 2, 3, \ldots, n)]^2 \, dV_1 \, dV_2 \, dV_3, \ldots, dV_n + e \sum_j z_j \vec{r}_j \quad \text{(III.22)}$$

where e is the electronic charge, $\vec{r}_i$ is the radius vector of the i-th electron, and z_j and $\vec{r}_j$ are, respectively, the charge of the j-th atomic nucleus and its radius vector.

Thus, knowing the wave function Ψ for a known or given structure and geometry of the molecule, it is possible to calculate the theoretical magnitude of its dipole moment in any possible form. A comparison of the calculated values with the experimental figures provides the possibility of deciding between one or another proposed structure and conformation.

It is obvious that in principle such an approach has enormous advantages over the vectorial additive scheme of calculation based on a set of empirically selected bond and group moments and enables all the effects not included in the framework of this scheme to be taken into account. However, the accuracy of dipole moments calculated by the quantum-mechanical method is seriously limited by the degree of approximation of the function Ψ to the true wave function of the molecule.

At the present time, the state of the methodical and computing apparatus of quantum chemistry is such that for complex molecules a fairly good approximation to the true wave functions can be obtained only by using empirical and semiempirical methods involving many different and not always theoretically based assumptions. These calculations require the a priori evaluation of a number of complex and not always accurately determined integrals which must be given in the form of parameters of the calculation. The final wave function and the electron distribution determined by it depend greatly on the selection of the initial parameters. Consequently, the experimental dipole moment is frequently used as a test to confirm the suitability of the parameters chosen, while the more important structural problem — the establishment of a structure in comparison with the calculated value of the moment — must be solved only when there are grounds for accepting the choice of the empirical parameters for calculation.

In the general case of molecules with unsaturated bonds, the calculation of the electronic structure is also carried out in the so-called σ, π approximation, i.e., on the assumption of the absence of a direct interaction between the σ and the π electrons. The σ-electron and π-electron distributions corresponding to this are calculated and the σ- and π-components of the total dipole moment μ_σ and μ_π are evaluated.

The calculation of the π-components of the moments of complex molecules is generally based either on a simple variant of the LCAO MO method in Hückel's approximation [103] or on its more rigorous semiempirical variants [104] of the self-consistent field (Pople) and configurational interaction (Pariser-Parr). One result of the calculation is the distribution of the effective π-electronic charges over the atoms. The dipole moment of a system of point electric charges q_i which is neutral as a whole is defined by the relation

$$\vec{\mu} = \sum_i q_i \vec{r}_i$$

(III.23)

the vector $\vec{\mu}$ not depending on the selection of the origin of coordinates. Thus, the problem consists in finding the value of $\vec{r}_i$ in a system of coordinates determined by considerations of convenience. We may illustrate the method of calculation with the case of the π-component of the dipole moment of imidazole (I).

In the molecular diagram (II) is shown the distribution of the π-charges on the atoms of imidazole obtained in Hückel's approximation with Streitwieser's parameters for the heteroatoms [103] and the auxiliary induction parameter h' = 0.1.

Considering the imidazole molecule as a regular pentagon with an internal angle of 108° ($\alpha = 36$, $\beta = 72°$) and a side of 1.38 Å and locating the origin of coordinates on the N_1 atom, we can define the vectors (in Å):

$$\vec{r}_2 = 1.38\,[(\sin \alpha)\,\vec{i} + (\cos \alpha)\,\vec{j}]$$
$$\vec{r}_3 = 1.38\,(2 \cos \alpha)\,\vec{j}$$
$$\vec{r}_4 = 1.38[(-\sin \beta)\,\vec{i} + (\cos \beta + 1)\,\vec{j}]$$
$$\vec{r}_5 = 1.38[(-\sin \beta)\,\vec{i} + (\cos \beta)\,\vec{j}]$$

(III.24)

where $\vec{i}$ and $\vec{j}$ are the unit vectors directed along the x and y axes, respectively.

On substituting the values of q_i and $\vec{r_i}$ in (III.23), we obtain the magnitude of the vector $\vec{\mu}_\pi$ while, since e = 4.8 · 10^{-10} CGSE and 1 Å = 10^{-8}cm, the value of $\vec{\mu}_\pi$ is calculated in Debye's: 10^{-18} CGSE = 1 D.

$$\vec{\mu}_\pi = -0.76\vec{i} + 2.93\vec{j}D \qquad (III.25)$$
$$|\mu_\pi| = 3.03\,D$$

The value obtained depends fundamentally on the selection of the parameters for calculation. Thus, without taking into account the auxiliary inductive parameter μ_π = 3.40 D, while μ_π calculated for the imidazole molecule from the π-electronic distribution in one of the variants of the MO method of the self-consistent field [124] is only 2.6 D.

At the present time there are no reliable methods for the experimental evaluation of the π-component of the dipole moment. The variants of the calculation of μ_π from the experimental value of the dipole moment proposed by Halverstadt and Kumler [105] and also by Kuhn [106] have, rather, a purely qualitative value. In view of this, the definitive checking of the degree of suitability of the parameters of the quantum-mechanical calculation for the reproduction of the dipole moment can be effected only on the basis of the calculated value of the overall dipole moment.

As already mentioned, to calculate the overall moment we must calculate the σ-component of the dipole moment of the molecule. We may refer to three of the approaches for the calculation of μ_σ used at the present time.

A number of authors, starting with the first paper on the quantum-mechanical calculation of dipole moments of complex molecules containing heteroatoms [107], have used a conventional vector scheme with the bond moments given in Table 8 to calculate μ_σ.

Another approach [11-13] is based on the theoretical calculation of the moments of all the σ-bonds of the molecule according to Mulliken and Coulson (see Chapter II) with the inclusion of homopolar and atomic dipoles. Its advantage is the possibility of calculating the moment of the unshared pairs of electrons, but this method is laborious and, as mentioned above, the use of the Slater atomic orbitals in the calculations is not always justified.

The most promising method for the calculation of the σ-component of the dipole moment is apparently that using the magnitudes of the σ-electronic charges on the atoms. The latter can be calculated by the general LCAO MO method (for reviews see [108-110]) or by means of the Del Re method mentioned above [19, 111], in which the orbitals localized in the individual bonds of the molecule are considered and the induction effects leading to a displacement of the electron cloud in the direction of the more electronegative atoms are taken into account. We may use the method to calculate the charge distribution on the atoms in saturated systems and in the σ-system of unsaturated molecules. The latter is achieved by suitable parametrization, implicitly taking into account the influence of the hybridization of the atoms in the molecule on their effective electronegativity [112]. The principle of the method is given in detail in the literature [19, 113]. Programs have been developed for the numerical realization of the method on electronic computers [114, 115].

The molecular diagram (III) gives the σ-electronic distribution in the molecule of imidazole calculated by the Del Re method with parameters taking into account the trigonal hybridization of the atoms [112].

III

Assuming that the length of the C—H bond is 1.08 Å, we obtain from (III.23) in D:

$$\vec{\mu}_\sigma = 0.74\vec{i} + 1.11\vec{j} \tag{III.26}$$

and by combining (III.25) and (III.26):

$$\vec{\mu} = \vec{\mu}_\pi + \vec{\mu}_\sigma = -0.02\vec{i} + 4.04\vec{j} \tag{III.27}$$

$$|\mu| = 4.04\ D$$

The calculated value agrees excellently with the experimental value of 3.99 D [116]. The use of the series of parameters mentioned above to calculate the π-electronic charges and the σ-

TABLE 16. Dipole Moments of Some Nitrogen–Containing Heterocyclic Compounds and Benzene Derivatives

Compound	μ_π, D	μ_σ, D	$\mu = \|\vec{\mu}_\pi + \vec{\mu}_\sigma\|$, D	μ_{exp}, D [48]
Pyridine *	0.86	1.44	2.30	2.21
Pyrimidine *	0.87	1.50	2.37	2.42
Pyrrole *	2.29	0.49	1.80	1.74
Indole *	2.56	0.53	2.05	2.05
Aniline *	1.11	0.40	1.51	1.53
Phenol *	1.31	1.65	1.46	1.55
4-Aminopyridine	2.11	1.83	3.94	3.95
9-Methyladenine	2.70	0.37	3.06	3.0 [119]
9-Methylpurine	3.53	0.73	4.26	4.3 [119]
1,3-Dimethyluracil	3.28	0.86	3.94	3.9 [119]
Caffeine	2.96	0.97	3.38	3.4
N-Methylpyrrole	2.32	0.40	1.92	1.94
Benzimidazole	3.46	1.35	4.02	4.03
Benzylideneaniline†	1.53	0.66	1.76	1.61

* The parameters of the heteroatoms for calculating the π-electronic charges
   were chosen from the dipole moments of the compounds given.
† The distribution of the π-charges was calculated by the Pariser–Parr method
   [120] with the inclusion of the configurational interaction.

charges by the Del Re method also leads to very good agreement between the calculated and experimental dipole moments in the case of a whole series of other imidazole derivatives [117].

Table 16, drawn up from the results of calculations [112, 118] gives other examples of the successful reproduction of the experimental values of the dipole moments of a number of aromatic nitrogen-containing heterocycles. As in the example described above, the calculation of the total moment was performed by the vector summation of the π-component calculated from the Hückel distribution of the π electrons and the σ-component evaluated by the Del Re method.

A similar approach has been used successfully for the calculation of the dipole moments of a number of derivatives of thiophene, tropone, thiotropone, and their isoelectronic analogs [121].

Thus, the selection of the parameters for calculating the π-electronic distribution from the dipole moments of some model compounds and the calculation of the σ-moment by the Del Re method may be used as a fairly reliable means for the theoretical calculation of the dipole moments of complex molecules.

Recently, quantum-mechanical methods have been developed which make it possible simultaneously to consider the σ and π-electrons of a fairly complex organic molecule within the framework of a single approximation. The simplest of these is the generalized Hückel method [125]. In principle, this method is analogous to the simple Hückel method, and its use for calculating dipole moments leads to very high values of the moments of organic compounds [126].

The SCF MO method developed by Pople, Santry, and Segal in the approximation of the complete neglect of differential overlapping (CNDO) is considerably more accurate. Thanks to the possibility of calculating the electronic populations of all the valence orbitals, we can calculate all the components of the over-all dipole moment (see Section 1 of Chapter II) and achieve very good agreement between the calculated and experimental dipole moments [129, 130]. At the present time, the use of this method is still restricted by the dimensions of the molecules of the compounds studied because of the limited possibilities of computers, but the accumulated experience of calculations already gives grounds for assuming that the method has exceptionally favorable prospects for the theoretical analysis of dipole moments and also for problems of the spatial structure of organic compounds. An important advantage of quantum-mechanical methods of calculation is the explanation by them of certain factors not included in the usual additive scheme: the polarity of nonalternant hydrocarbons, the difference in the dipole moments of the s-cis and s-trans isomers with a single polar bond as, for example, in acrolein, etc. The quantum-mechanical method of calculation enables the influence of certain fine structural effects (the hydrogen bond [94, 112, 122], the contribution of vacant d-orbitals [123], etc.) on the dipole moments of the molecule to be evaluated. It is particularly important that quantum-mechanical methods of calculating dipole moments are not limited by the type of compounds studied and by additive relationships as is characteristic for the vectorial scheme and are also applicable to calculations of dipole moments of molecules in excited states.

It may be considered that in future the development of the calculating apparatus of the dipole moment method will be directly connected with the perfection and adaptation of quantum-mechanical calculations of electronic structure of molecules.

Literature Cited

1. J. J. Thomson, Phil. Mag., 46:513 (1923).
2. C. A. Coulson, Trans. Faraday Soc., 38:433 (1942).
3. R. S. Mulliken, J. Chem. Phys., 46:539 (1949).
4. H. Hartmann, Naturforsch., 2A:489 (1947); 3:47 (1948).
5. W. L. G. Gent, Quart. Revs., 2:383 (1947).
6. A. D. Walsch, Discuss. Faraday Soc., 2:18 (1947).
7. C. Coulson, Valence, Oxford University Press, 2nd Ed. (1961).
8. W. C. Hamilton, J. Chem. Phys., 26:345 (1957).
9. H. Glazer and H. Reiss, J. Chem. Phys., 23:937 (1955).
10. C. A. Coulson, Proc. Roy. Soc., A207:63 (1951).
11. J. H. Gibbs, J. Phys. Chem., 59:644 (1955).
12. H. M. Hameka and A. M. Liquori, Mol. Phys., 1:9 (1958).
13. J. B. Lounsbury, J. Phys. Chem., 67:721 (1963).
14. F. O. Ellison and H. Schull, J. Chem. Phys., 28:2348 (1958).
15. L. Burnell and C. A. Coulson, Trans. Faraday Soc., 53:403 (1957).
16. A. Julg and M. Bonnet, J. Chim. Phys., 60:742 (1963).
17. W. Seiffert, H. Zimmerman, and G. Scheibe, Angew. Chem., 74:249 (1962).
18. P. Coppens and F. L. Hirschfeld, Israel J. Chem., 2:117 (1954).
19. G. Del Re, J. Chem. Soc., 1958:4031.
20. L. Pauling, The Nature of the Chemical Bond, Cornell University Press, New York, 3rd Ed. (1960).
21. S. S. Batsanov, The Electronegativity of the Elements and the Chemical Bond, Izd. Sib. Otd. Akad. Nauk SSSR (1962), pp. 67-85.
22. R. Ferreira, Trans. Faraday Soc., 59:1064 (1963).
23. J. Hinze, M. A. Whitehead, and H. H. Jaffe, J. Am. Chem. Soc., 85:148 (1963).
24. G. V. Vykov, Electronic Charges of Bonds in Organic Compounds, Izd. Akad. Nauk SSSR (1960), Chap. 3.
25. R. L. Cook and A. P. Mills, J. Phys. Chem., 65:252 (1961).
26. M. D. Newton, F. P. Boer, and W. N. Lipscomb, J. Am. Chem. Soc., 88:2361, 2367 (1966).
27. D. F. Hornig and D. C. McKean, J. Phys. Chem., 59:1133 (1955).
28. L. A. Gribov, Theory of the Intensities in the Infrared Spectra of Polyatomic Molecules, Izd. Akad. Nauk SSSR (1963).
29. M. V. Vol'kenshtein, M. A. El'yashevich, and B. I. Stepanov, Vibrations of Molecules, Gos. Izd. Tekh. i Teoret. Lit., Vol. 2 (1949).
30. R. C. Golike, J. M. Mills, W. B. Person, and B. L. Crawford, J. Chem. Phys., 25: 1266 (1956).
31. K. Higasi, H. Baba, and A. Rembaum, Organic Quantum Chemistry, Interscience, New York (1965), Chap. 8.

32. O. Theimer and R. Theimer, J. Mol. Spectr., 8:236 (1962).
33. L. M. Sverdlov, Optika i Spektroskopiya, 4:697 (1958); 8:594 (1960).
34. R. H. Cole and A. J. Michell, Spectrochim. Acta, 20:739 (1964).
35. M. A. Kovner and B. N. Snegirev, Optika i Spektroskopiya, 10:328 (1961).
36. L. A. Gribov and E. M. Popov, Dokl. Akad. Nauk SSSR, 145:761 (1962).
37. R. E. Hiller and J. W. Straley, J. Mol. Spectr., 5:24 (1960).
38. L. M. Sverdlov and Yu. V. Klochkovskii, Optika i Spektroskopiya, 17:466 (1964).
39. E. M. Popov and V. A. Shlyapochnikov, Optika i Spektroskopiya, 14:779 (1963).
40. R. Rollefson and R. Havens, Phys. Rev., 57:710 (1940).
41. A. J. Petro, J. Am. Chem. Soc., 80:4230 (1958).
42. R. L. Kelly, R. Rollefson, and B. S. Schurin, J. Chem. Phys., 19:1595 (1951).
43. H. Bent, Chem. Rev., 61:275 (1961).
44. A. Eucken and L. Meyer, Phys. Z., 30:397 (1929).
45. C. P. Smyth, J. Am. Chem. Soc., 60:183 (1938).
46. J. W. Smith, Electric Dipole Moments, Butterworths, London (1955).
47. C. P. Smyth, Dielectric Behavior and Structure, McGraw-Hill, New York (1955).
48. O. A. Osipov and V. I. Minkin, Handbook on Dipole Moments, Izd. Vysshaya shkola (1965).
49. Tables of Interatomic Distances and Configuration in Molecules and Ions, Spec. Publ. No. 11, Ch lical Society, London (1958).
50. H. Lumbroso and C. Audrien, Bull. Soc. Chim. France, 1966:3201.
51. D. C. McKean, J. Am. Chem. Soc., 77:2260 (1955).
52. R. S. Holland and C. P. Smyth, J. Am. Chem. Soc., 77:268 (1955).
53. J. W. Williams, Phys. Z., 29:683 (1928).
54. R. J. B. Marsden and L. E. Sutton, J. Chem. Soc., 1936:599.
55. N. Leonard and L. E. Sutton, J. Am. Chem. Soc., 70:1564 (1948).
56. B. Eda and K. Ito, Bull. Chem. Soc. Japan, 29:524 (1956); 30:164 (1957).
57. A. E. Lutskii and V. V. Dorofeev, Zh. Fiz. Khim., 33:331 (1959).
58. A. E. Lutskii, E. M. Obukhova, and B. P. Kondratenko, Zh. Fiz. Khim., 37:1270 (1963).
59. A. E. Lutskii, V. T. Alekseeva, and B. P. Kondratenko, Zh. Fiz. Khim., 35:1706 (1961).
60. W. A. Sheppard, J. Am. Chem. Soc., 85:1314 (1963).
61. J. W. Smith, J. Chem. Soc., 1961:81.
62. A. E. Lutskii and B. P. Kondratenko, Zh. Fiz. Khim., 33:2017 (1959).
63. A. E. Lutskii, L. M. Yagupol'skii, and E. M. Obukhova, Zh. Obshch. Khim., 34: 2641 (1964).
64. S. Millefiori and A. Foffani, Tetrahedron, 22:803 (1966).
65. V. Baliah and M. Uma, Tetrahedron, 19:455 (1963).
66. A. R. Katritzky, E. W. Randall, and L. E. Sutton, J. Chem. Soc., 1957:1769.
67. H. Lumbroso and D. M. Bertin, Bull. Soc. Chim. France, 1966:532.
68. E. Eliel, Stereochemistry of Carbon Compounds, McGraw-Hill, New York (1962).
69. E. J. Corey and R. A. Sneen, J. Am. Chem. Soc., 77:2505 (1955).
70. C. F. Wilcox, J. Am. Chem. Soc., 82:414 (1960).
71. T. Henshall, J. Chem. Educ., 43:600 (1966).
72. L. Scott and A. McKenzie, J. Chem. Educ., 43:27 (1966).

73. E. L. Eliel, N. L. Allinger, S. J. Angyal, and G. A. Morrison, Conformational Analysis, Interscience, New York (1965), p. 454.

74. O. Fuchs, Z. Phys. Chem., B14:339 (1931).

75. L. Tiganik, Z. Phys. Chem., B14:135 (1932).

76. J. W. Williams, Z. Phys. Chem., A138:75 (1928).

77. H. Eyring, Phys. Rev., 39:746 (1932).

78. H. Lumbroso and D. Lauransan, Bull. Soc. Chim. France, 1959:513.

79. P. Trunei, Ann. Chim., 12:93 (1939).

80. C. T. Zahn, Phys. Z., 33:400 (1932).

81. T. S. Gilman, J. Am. Chem. Soc., 88:1861 (1966).

82. C. F. Wilcox, J. G. Zajacek, and M. F. Wilcox, J. Org. Chem., 30:2621 (1965).

83. P. Hirsjarvi and H. Krieger, Suomen Kemistislechti, B37:140 (1964).

84. D. D. Tanner and T. S. Gilman, J. Am. Chem. Soc., 85:2892 (1963).

85. O. Exner and V. Jehlicka, Collection Czech. Chem. Commun., 30:639 (1965).

86. O. Exner, Collection Czech. Chem. Commun., 30:652 (1965).

87. O. Exner, Chem. Listy, 60:1047 (1966).

88. A. N. Vereshchagin and S. G. Vul'fson, Teoret. Éksperim. Khim., Akad. Nauk UkrSSR, 1:305 (1965).

89. E. Gei [E. Gey], S. S. Yarovoi, and V. M. Tatevskii, Vestn. Mosk. Univ., Ser. Khim., No. 1:9; No. 3:15; No. 4:3 (1965).

90. E. Gey, Z. Chem., 6:281 (1966).

91. E. Gey, Theor. Chim. Acta, 5:187 (1965).

92. E. Gey, Z. Chem., 7:351 (1967).

93. V. M. Tatevskii, The Chemical Structure of Hydrocarbons and Regularities in Their Physicochemical Properties, Izd. Mosk. Univ. (1953).

94. V. I. Minkin, Yu. A. Zhdanov, I. D. Sadekov, and A. D. Garnovskii, Dokl. Akad. Nauk SSSR, 162:108 (1965).

95. V. I. Minkin, O. A. Osipov, and V. A. Kogan, Dokl. Akad. Nauk SSSR, 145:336 (1962).

96. V. I. Minkin, Yu. A. Zhdanov, A. D. Garnovskii, and I. D. Sadekov, Zh. Fiz. Khim., 40:657 (1966).

97. L. M. Nazarova and Ya. K. Syrkin, Izv. Akad. Nauk SSSR, Ser. Khim., 35,1949; Zh. Obshch. Khim., 23:478 (1953).

98. T. Shimozawa, Bull. Chem. Soc. Japan, 38:1046 (1965).

99. S. Walker, in: Physical Methods in the Chemistry of Heterocyclic Compounds, ed. A. P. Katritzky, Academic Press (1963).

100. D. J. W. Bullock and C. W. N. Cumper, J. Chem. Soc., 1965:5311.

101. I. Mazheika, L. Avota, G. Sokolov, and S. A. Giller, Zh. Obshch. Khim., 34:3380 (1964).

102. S. A. Giller, I. B. Mazheika, and I. I. Grandberg, Khim. Geterotsikl. Soedin., 1965:103, 107.

103. A. Streitwieser, Molecular Orbital Theory for Organic Chemists, Wiley, New York (1961).

104. R. G. Parr, The Quantum Theory of Molecular Electronic Structure, Benjamin, New York (1963).

105. I. F. Halverstadt and W. D. Kumler, J. Am. Chem. Soc., 64:2988 (1942).

106. H. Kuhn, Die Methode des Electronengases. Vorträge am Intern. Ferienkurs die Theorie der π-Elektronensystem [The Electron Gas Method. Lectures at the International Holiday Course on the Theory of the π-Electron System], Berlin-Heidelberg (1963).

107. L. E. Orgel, T. L. Cottrell, W. Dick, and L. E. Sutton, Trans. Faraday Soc., 47: 113 (1951).

108. G. Klopman, Tetrahedron, Suppl. 2:111 (1963).

109. K. Fukui, in: Molecular Orbitals in Physics, Chemistry and Biology, A Tribute to R. S. Mulliken, ed. P.-O. Löwdin and B. Pullman, Academic Press, New York (1964), p. 513.

110. N. D. Sokolov, Usp. Khim., 36:2195 (1967).

111. G. Del Re, B. Pullman, and T. Yonezawa, Biochim. Biophys. Acta, 75:153(1963).

112. H. Berthod and A. Pullman, J. Chim. Phys., 62:942 (1965).

113. L. B. Kier, Tetrahedron Letters, 1965:3273.

114. Yu. A. Ostroumov, Programme for Calculating Molecules by the Hückel Molecular Orbital Method on an M-20 Electronic Computer, Izd. Rostov-on-Don Univ. (1965).

115. Yu. A. Kruglyak and I. S. Yashchenko, Teoret. Éksperim. Khim., Akad. Nauk UkrSSR, 2:840 (1966).

116. V. I. Minkin, O. A. Osipov, A. D. Garnovskii, and A. M. Simonov, Zh. Fiz. Khim., 36:469 (1962).

117. V. I. Minkin, A. D. Garnovskii, L. S. Éfros, and Yu. A. Zhdanov, The Chemistry of Five-Membered Nitrogen-Containing Heterocycles. Abstracts of Papers at the Second All-Union Conference, Rostov-on-Don (1966), p. 78.

118. H. Berthod and A. Pullman, Biopolymers, 2:483 (1964).

119. H. de Voe and I. Tinoko, J. Mol. Biol., 4:500 (1962).

120. V. I. Minkin, Yu. A. Zhdanov, E. A. Medyantzeva, and Yu. A. Ostroumov, Tetrahedron, 23:3651 (1967).

121. A. Melhorn, Chem. Listy, 61:32 (1967).

122. H. Berthod and A. Pullman, Compt. Rend., 259:2711 (1964).

123. O. Sovers and W. Kauzmann, J. Chem. Phys., 38:813 (1963).

124. R. D. Brown and M. L. Heffernan, Australian J. Chem., 12:543 (1959).

125. R. Hofmann, J. Chem. Phys., 39:1397 (1963).

126. W. Adam and A. Grimison, Theor. Chim. Acta, 7:342 (1967).

127. J. A. Pople, D. P. Santry, and G. A. Segal, J. Chem. Phys., 43:129, 136 (1965).

128. J. A. Pople and G. A. Segal, J. Chem. Phys., 44:3289 (1966).

129. J. A. Pople and M. Gordon, J. Am. Chem. Soc., 89:4253 (1967).

130. J. E. Bloor and D. L. Breen, J. Phys. Chem., 72:716 (1968).

Chapter IV

Dipole Moments and the Stereochemistry of Organic Compounds

1. The Dipole Moment and the Symmetry of the Molecule

Centrosymmetrical and Noncentrosymmetrical Compounds. The simplest application of the method of dipole moments to the study of the spatial structure of molecules is the choice, based on the experimental value of the dipole moment of a compound with an unknown structure, between two or more structures one of which possesses a center of symmetry. The dipole moment of such a structure must be zero, since the moments of all the polar groups present in the molecule compensate one another as, for example, in compounds (I)–(III):

| I | II | III |
| $\mu = 0 \; (CCl_4)$ [1] | $\mu = 0$ [2] | $\mu = 0$ [3] |

The fact that the dipole moments of diindenyliron and diindenylruthenium [4] are zero shows that they exist in solutions in

the form of the stable centrosymmetrical trans configuration (IV),
while in the crystalline state, according to X-ray data [5], the polar
cis configurations (V) are preferred.

Me = Fe, Ru

IV   V

Metal chelates of type (VI) exist in solutions in the form of
an equilibrium mixture of the centrosymmetrical planar (VII) and
the tetrahedral (VIII) relative to the coordination center of the conformations.

The relative amounts of the conformers mentioned depend on
the spatial structure of the aryl residue and can be evaluated from
the magnitude of the dipole moments. Thus, the dipole moments of
the mesidine complexes (Ar = $2,4,6\text{-}C_6H_2(CH_3)_3$) of nickel and copper (in benzene) are 0.41 and 0.64 D, respectively. The low values
of the moments show that only a small proportion of conformation
(VIII) is present and the centrosymmetrical structure (VII) predominates. On the other hand, for the complexes of nickel and
copper with Ar = C_6H_5, values of 2.68 [6] and 2.70 D [7] have been
found, indicating a high concentration of the tetrahedral conformations (VIII). A number of other examples of the investigation of
the configurations of the chelate center of internally complex compounds with organic liquids has been given in a review [8].

In many cases, the presence or absence of polarity in a newly
obtained compound can be used successfully for its identification.
Thus, the dipole moment of the dimer of isoamylideneaniline of
2.28 D permits an easy decision between the two alternative structures (IX) and (X) in favor of the latter [9]

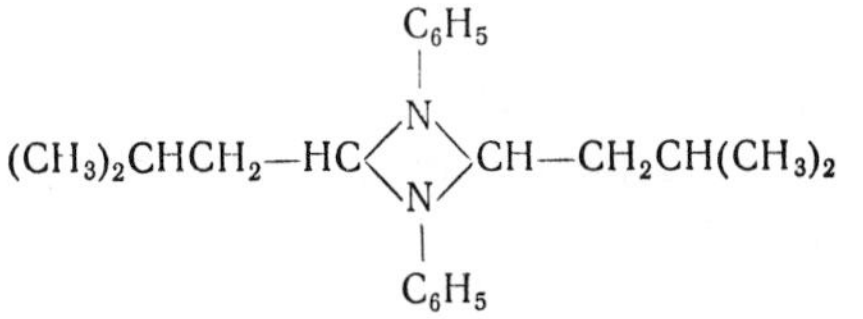

$$C_6H_5$$
$$|$$
$$N$$
$$(CH_3)_2CHCH_2-HC \Big\langle \quad \Big\rangle CH-CH_2CH(CH_3)_2$$
$$N$$
$$|$$
$$C_6H_5$$
IX

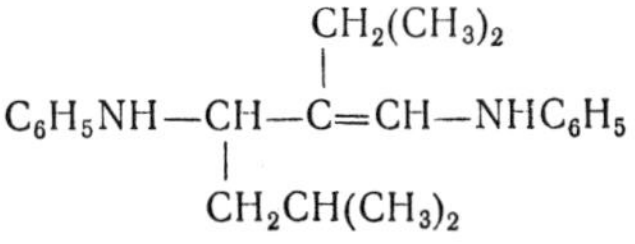

$$CH_2(CH_3)_2$$
$$|$$
$$C_6H_5NH-CH-C=CH-NHC_6H_5$$
$$|$$
$$CH_2CH(CH_3)_2$$
X

Similarly, the fact that the dimethylbenzobisthiazole obtained by Kiprianov and Mikhailenko [10], the structure of which, from the reaction involved, could be either the centrosymmetrical (XI) or (XII), has a dipole moment (2.7 D) definitely indicates the angular structure (XII) in spite of the assignments from other studies.

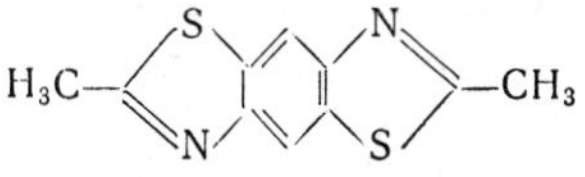

$$H_3C- \quad -CH_3$$
XI

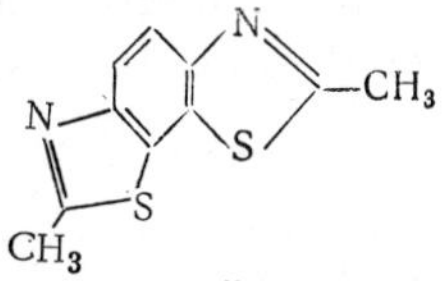

$$-CH_3$$
$$CH_3$$
XII

The detection of polarity in the anthrenes (XIII) and in the trimer of phosphonitrile chloride (XIV) also contradicts the centrosymmetrical configurations assigned to them.

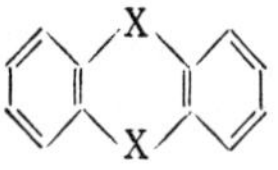

XIII

X	μ, D
O	0.64 [11]
S	1.57 [12]
Se	1.41 [12]

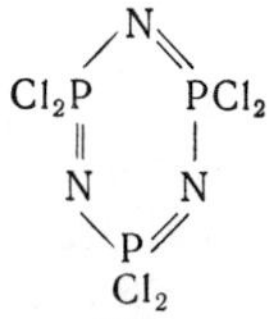

$$Cl_2P \quad PCl_2$$
XIV
$$\mu = 0.93D \ [13]$$

In actual fact, as X-ray structural studies show, the molecules of (XIII) are bent about the line X–X [14], while (XIV) undergoes considerable distortions in crystals [15]. In solutions, moreover, the presence of several conformations (including polar ones) of compounds (XIII) and (XIV) is possible, which must lead to an experimentally observable dipole moment.

In many cases, particularly when the observed dipole moment is small, a decision on the presence or absence of a deviation of the molecular structure from the centrosymmetrical structure can be made only after a direct determination of the atomic or deformation polarization. Thus, Hampson [16] has found that the difference between the total and electronic polarizations of diphenylmercury and a number of p, p'-disubstituted (CH_3, F, Cl, Br) derivatives is 4-20 cm³ and he ascribed it to orientation polarization. It followed from this that diphenylmercury and its derivatives possess a dipole moment (0.4-1.0 D), i.e., the C—Hg—C angle is less than 180°. However, Coop and Sutton [17] suggested that this difference was due to an anomalous atomic polarization and that the molecules of diarylmercuries are nonpolar and have a linear structure.

Only the special investigation of Wright et al. [18], in which the deformation polarizations were determined directly, showed that the difference between the total and deformation polarization has an orientation origin, and compounds of the type of the diarylmercuries consequently possess an angular structure.

Other Types of Symmetry Leading to the Absence of Polarity. A centrosymmetrical structure is a sufficient but not a necessary condition for the nonpolarity of a molecule. A more general formulation follows from expression (III.1) and requires only that the vectorial sum of the moments of all the polar groups present in the molecule be zero. This can be achieved in other types of symmetry having no center of inversion among the elements of symmetry, for example in the tetrahedral type. Thus, the nonpolarity of methane, carbon tetrachloride, and tin tetrachloride shows the tetrahedral configuration of the bonds about the central atom and the absence of dipole moments in alkanes (or the existence of only very small ones) shows very small deviations from tetrahedrality.

In the general case, the polarity of any particular compound is determined not by the symmetry of the whole molecule but by the symmetry of the mutual orientation of the polar groups. Thus, phenanthrene (XV) has no dipole moment [2], although it has only C_{2v} symmetry, because the arrangement of the vectors of the moments of the C—H polar groups is centrosymmetrical.

XV   XVI

XVII

In formulas (XVI) and (XVII), the asterisks mark overlapping CH spheres.

The same situation should apparently be observed for 3,4-benzophenanthrene (XVI); however, this is found to have a dipole moment of 0.7 D [19]. The reason for this is that the molecule of (XVI) is not in fact planar because of the spatial overloading which disturbs the C_{2v} molecular symmetry and the centrosymmetrical orientation of the dipoles of the C −H bonds [20]. On the other hand, tetrabenzonaphthalene (XVII), the molecule of which is not planar for the same reasons as (XVI) has a zero dipole moment. Consequently, the deviation from planarity of the molecule of (XVII) does not disturb its centrosymmetry.

<u>Dipole Moments and Position Isomerism</u>. The dipole moments of the ortho, meta, and para isomers of disubstituted benzenes differ fundamentally from one another (Table 21). In the general case, the same situation is preserved for isomers of polysubstituted benzenes. Thus, dipole moment data can be used successfully to identify position isomers.

At the present time, this problem is no longer an urgent one for compounds of the benzene series, but in a number of polynuclear and heterocyclic compounds the determination of the dipole moment is frequently the most convenient method of deciding between two or more proposed structures for isomers [21].

We may mention here an interesting example from the field of organoboron compounds with a polyhedral structure − the carboranes. The magnitudes of the dipole moments of the three isomers of dicarbadecaborane (XVIII), $C_2H_2B_{10}H_{10}$, must, as follows from symmetry considerations, decrease in the sequence from the ortho isomer to the para isomer, the latter being nonpolar.

XVIII

1,2-CH, o-carborane, μ = 4.53 D
1,3-CH, m-carborane, μ = 2.85 D
1,4-CH, p-carborane, μ = 0

The experimental values of the dipole moments [22] confirm these assumptions and agree with the assignment of the isomers based on the methods of their production.

Determination of Structural Parameters. It is obvious that relation (III.5) can be used to find the angle between two dipoles if their group moments and the resultant dipole moment are known. Formula (III.5) can therefore be used in the calculation of valence angles. For example, the $C-O-C$ angle in di(p-bromophenyl) ether (XIX) (μ = 0.86 D) in the gas phase can be calculated from the moment of diphenyl ether (XX) (μ = 1.14 D) and the $Ar-Br$ moment (see Table 11).

XIX   XX

on the assumption that the $C-O-C$ angle is the same for the two compounds:

$$\mu_{XX}^2 = 2\mu_{Ar-O}^2 + 2\mu_{Ar-O}^2 \cos \varphi$$

$$\mu_{XIX}^2 = 2\left(\mu_{Ar-Br} - \mu_{Ar-O}\right)^2 + 2\left(\mu_{Ar-Br} - \mu_{Ar-O}\right) \cos \varphi \qquad (IV.1)$$

By deducting μ_{Ar-O} from equations (IV.1), we obtain φ = 105°. In this calculation the additional conjugation between the ether oxygen and the bromine was not taken into account (see Chapter V). Leonard and Sutton [23] took effects of this type into account and substantiated their preference for the use of the dipole moments of difluoro derivatives to calculate valence angles. The value of the valence angles calculated by them for several analogs of diphenyl ether, together with the results of electron-diffraction and X-ray structural determinations, are given in Table 17.

It must, however, be mentioned that such good agreement as in Table 17 of the values of the valence angles calculated from dipole moments and those found on the basis of structural investigations is obtained in by no means all cases. The values of the angles calculated from the dipole moments are extremely sensitive to the selection of the values of the group moments and to the introduction of errors due to mutual influences (Chapter V). In view of this, data on dipole moments are generally suitable only for rough qualitative evaluations of valence angles.

TABLE 17. Valence Angles of Ar−X−Ar (in Degrees)

X	Calculated from the dipole moments taking the interaction moments into account	Structural data	X	Calculated from the dipole moments taking the interaction moments into account	Structural data
O	115—117	118 [24]	SO_2	93—106	100 [25]
S	103—106	109.5 [25]	CO	128—132	127 [26]

2. Geometrical cis-trans Isomerism

Relative to Double Bonds

Isomerism Relative to a C=C Bond. The determination of the geometrical configuration of isomers due to the presence of a C=C bond in organic compounds is one of the earliest applications of the dipole moment method to the study of structural problems.

In the majority of cases, dipole moments enable an unambiguous choice to be made between cis and trans isomers on the basis of purely qualitative comparisons. Thus, the dipole moment of trans-difluoroethylene (I) must be zero and that of the cis-isomer (II) appreciably different from zero, which is in agreement with the experimental results [27].

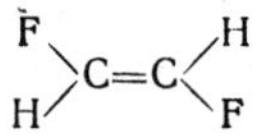

I
$\mu = 0$

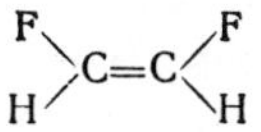

II
$\mu = 2.42D$

The assignment of the cis-trans isomers is equally obvious to many other cases, for example:

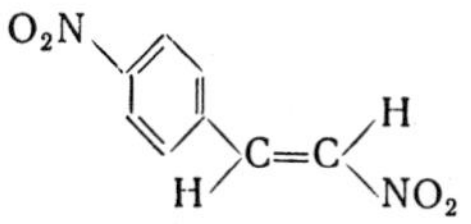

III
$\mu = 0.50\ D$ [28]

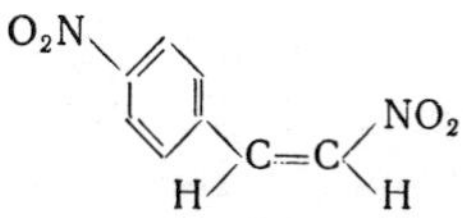

IV
$\mu = 7.38\ D$ [28]

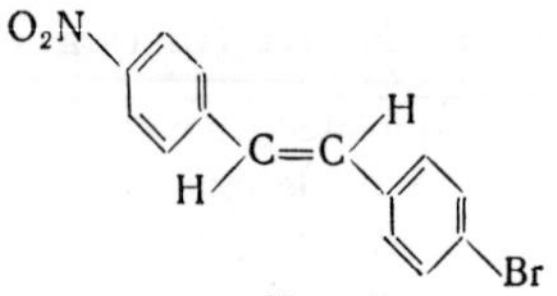

V

$\mu = 3.11\ D$ [29]

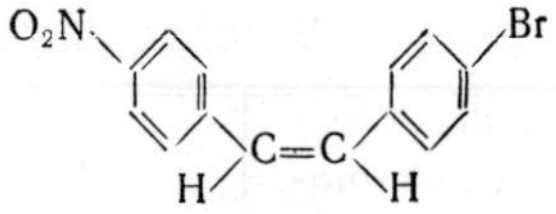

VI

$\mu = 4.52\ D$ [29]

It is clear from the examples given that the determination of cis-trans configurations from the dipole moments of the isomers is easy to carry out if the molecule contains two regular polar groups. As a rule, the presence of only one polar substituent is insufficient for judging the configuration on the basis of dipole moments alone. The presence of a large number of polar groups in the molecule also frequently complicates the choice of configuration. For example, the assignment of the isomers of dibromofluoroethylene

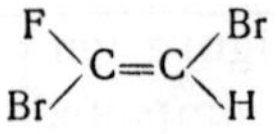

VII

$\mu = 1.36\ D$

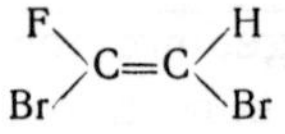

VIII

$\mu = 1.20\ D$

is based [30] on the assumption that the dipole moment of (VII) must be very close to the moment of vinyl fluoride (1.5 D [27]) and that of (VIII) less than that of vinyl bromide (1.41 D [31]). In this case, however, the difference in the dipole moments is too small for such an assignment to be regarded as unambiguous.

The choice of configuration can be carried out not only on the basis of the dipole moment but also by determining the direction of its vector. Thus, from the dipole moments of the dimers of ketene (IX) and of dimethylketene (X)

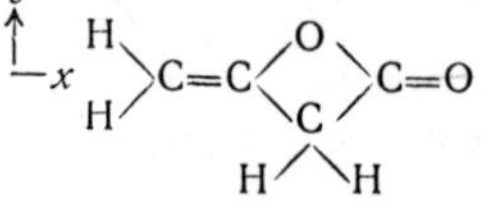

IX

$\mu = 3.23\ D$ [32]

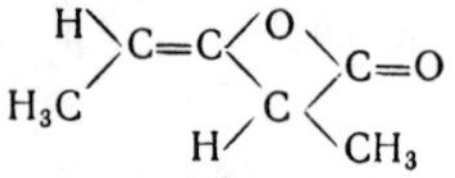

X

$\mu = 3.30\ D$ [33]

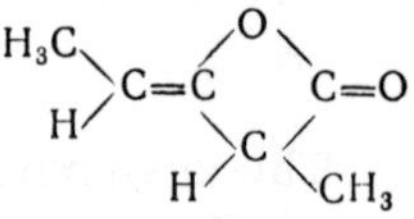

XI

and the usual values of the group moments it is possible to calculate (see Chapter III) the direction of the vector of the dipole moment in the trans (X) and cis (XI) configurations. For compound (X) the angle of inclination of the vector of the dipole moment to the X axis is 127° and for (XI) it is 3°. Since the largest group moment in (X) and (XI) is possessed by the C=O group and the direction of the total moment must be close to the direction of the C=O dipole, it is clear that the choice between (X) and (XI) must be made in favor of (XI) [33].

Isomerism Relative to a N=N Bond. The dipole moment of trans-azobenzene (XII) is zero. The UV irradiation of compound (XII) gives rise to the cis isomer (XIII) with a high dipole moment [34].

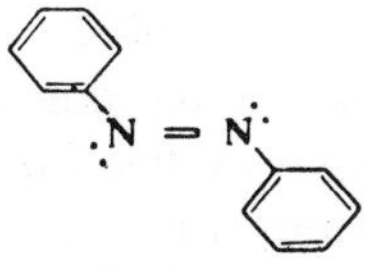

XII

$\mu = 0$

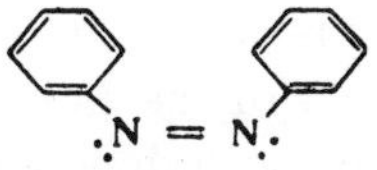

XIII

$\mu = 3.0\ D$

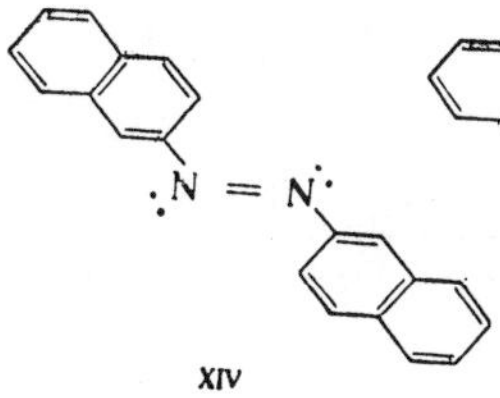

XIV

$\mu = 0$

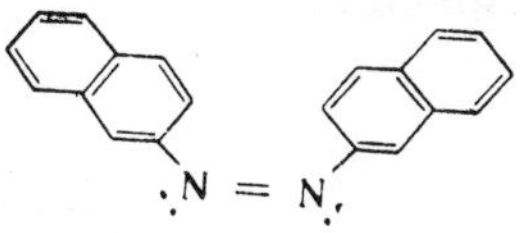

XV

$\mu = 3.0\ D$

Identical values of the dipole moments have also been found for the cis and trans isomers of 2,2'-azonaphthalene [35].

In the aliphatic series, the determination of the dipole moments of azomethane [36] (XVI) and 1-azobis-1-phenylpropane [37] (XVII) easily enables their trans configurations to be established.* This assignment is facilitated by a comparison with model compounds of type (XVIII) having a constrained cis structure [37].

$H_3C{-}N{=}N{-}CH_3$

XVI

$\mu = 0$

$C_6H_5(C_2H_5)CH{-}N{=}N{-}CH(C_2H_5)C_6H_5$

XVII

$\mu = 0.50\ D$

* The small dipole moment of compound (XVII) is the consequence of an inaccurate allowance for the atomic polarization or of contamination with a small amount of the cis form.

$$
\begin{array}{c}
\diagup (CH_2)_n \diagdown \\
C_6H_5-CH \qquad CH-C_6H_5 \\
\diagdown N\!=\!\!=\!\!N \diagup
\end{array}
$$

XVIII

$n = 1,\ \mu = 2.23\ D$

$n = 3,\ \mu = 2.42\ D$

$n = 4,\ \mu = 2.33\ D$

The cis isomers of azo compounds are unstable and are converted into the trans forms on heating. Since the dipole moments (molar polarization) of the isomers differ considerably from one another, a study of their dependence on the time during the thermal cis-trans conversion is a convenient method for investigating the kinetics of the isomerization reaction. If the molar polarizations of the pure trans and cis isomers, P_{trans} and P_{cis}, respectively, are known, the content of the cis isomer X at time t can be calculated from the formula

$$
X = \frac{P_t - P_{trans}}{P_{cis} - P_{trans}} \tag{IV.2}
$$

where P_t is the molar polarization of the mixture of isomers.

Similar investigations have been made for a series of azobenzenes [38], azonaphthalenes [35], and diazonium cyanides [39]. It was established that the reaction follows the first-order kinetic law. The calculated times of half-tranformation and the activation energy agree well with the results obtained by other methods.

Isomerism Relative to a C=N Bond. In contrast to azo compounds, the cis isomers of azomethines cannot be isolated at ordinary temperatures and these compounds always exist in the trans form. This is confirmed by the coincidence (within the limits of accuracy of the measurements) of the dipole moments of benzylideneaniline (XIX) and p,p'disubstituted derivatives (XX). It is obvious that such coincidence is possible only for the trans forms in which the vectors of the group moments of the similar para substituents have opposite directions.

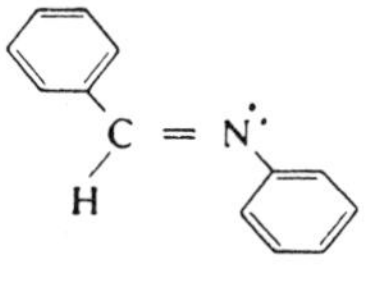

XIX   XX   XXI

$\mu = 1.61\ D$ [40]   R = Cl, $\mu = 1.56\ D$ [41] $\mu = 1.50\ D$ [41]

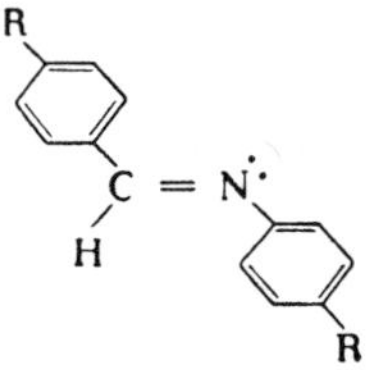

R = CH$_3$, $\mu = 1.58\ D$ [42]

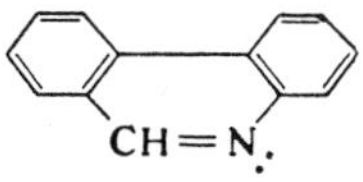

The approximate equivalence of the dipole moments of the unsubstituted and substituted compounds in this case is a decisive argument in favor of the trans configuration (XIX), since the dipole moments of the cis and trans forms calculated from the bond moments scarcely differ from one another [41], and the moment of phenanthridine (XXI), which can to some extent serve as an analog of the cis isomer, is also close to the moment of (XIX).

The N-oxides of the azomethines (the nitrones) can exist in two forms. If the molecule of the nitrone contains no other polar group besides the N → O group, measurement of the dipole moments easily enables the configuration of the compound to be determined as, for example, in the following cases:

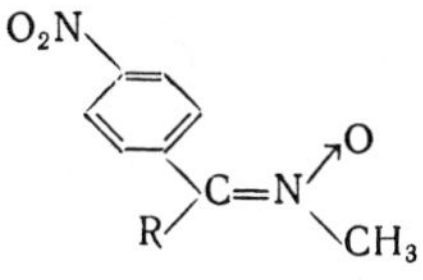

XXII

R = H, $\mu = 6.20\ D$ [43]

R = C$_6$H$_5$, $\mu = 6.60\ D$ [44]

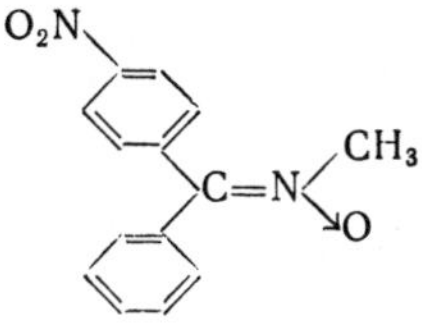

XXIII

$\mu = 1.09\ D$ [44]

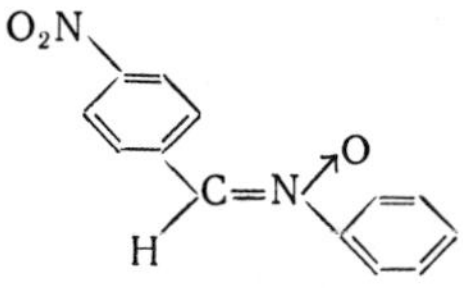

XXIV

$\mu = 6.32\ D$ [43]

An analogous example is the determination of the configuration of the N-chloroimines [45]

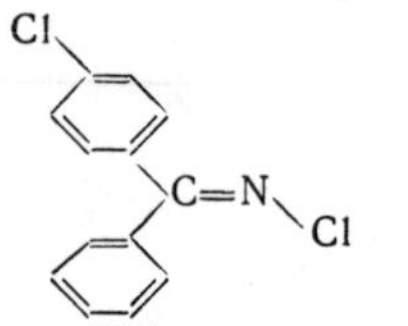

XXV
$\mu = 2.47\ D$

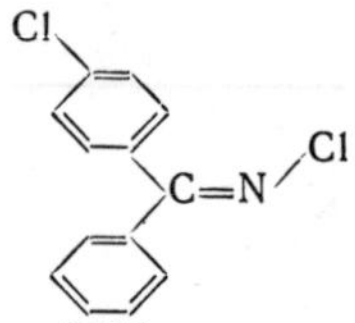

XXVI
$\mu = 2.67\ D$

The small differences in the moments of the cis and trans isomers are explained by the small moment of the N—Cl bond [21].

Another series of cases of isomerism relative to the C=N bond was considered by Exner [46] who made use of the graphical method of analyzing dipole moments that he had proposed (Chapter III). In agreement with the results of X-ray structural studies, he established the cis configuration of formamidoxime (XXVII), the trans structure of hydroxamoyl chloride (XXVIII), and so on.

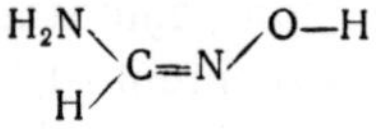

XXVII

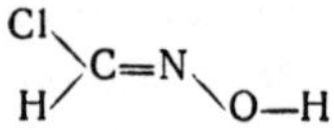

XXVIII

Finally, we may mention the interesting case of the stereo-isomerism of benzoquinone dianil:

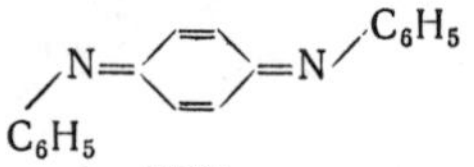

XXIX
$\mu = 0$

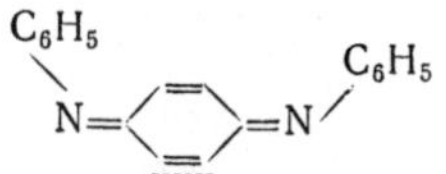

XXX
$\mu \neq 0$

The measured dipole moment of 1.67 D clearly shows the predominance of the cis arrangement of the phenyl groups, as in structure (XXX) [47].

Isomerism Relative to Other Double Bonds. The configuration of two isomers of oxythiobenzoyl chloride that had been synthesized were established by means of dipole moments [48]:

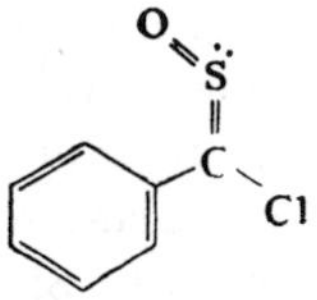

XXXI

$\mu = 2.63\ D$

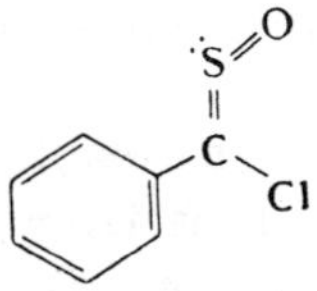

XXXII

$\mu = 3.97\ D$

It is clear that the dipole moment of the trans form (XXXI) will be lower than that of the cis form (XXXII) since the moments of the S=O and C−Cl bonds have their negative poles on the oxygen and the chlorine, respectively.

An interesting case of geometrical isomerism is found in the monoaminoboranes of type (XXXIII):

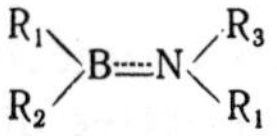

XXXIII

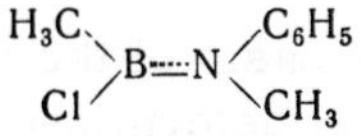

XXXIV

$\mu_{calc} = 2.32\ D$

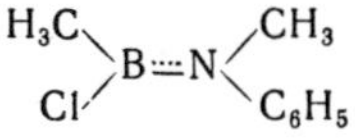

XXXV

$\mu_{calc} = 2.80\ D$

In addition to the σ-bond between the nitrogen and the boron there is an additional coordination bond through the filling of the vacant 2p orbital of the boron with the 2p electrons of the nitrogen. As a result, the B−N bond is similar to a double bond, and the aminoboranes (XXXIII) can exist as cis and trans isomers. Becker and Diehl [49] determined the dipole moments of a series of monoaminoboranes and from the results obtained they calculated the theoretical values of the moments of the cis and trans isomers (XXXIV) and (XXXV). From these values and the experimental figure of 2.37 D they calculated from (III.21) that in benzene at 25°C the equilibrium mixture (XXXIV)⇌(XXXV) contained about 90% of the trans isomer (XXXIV).

3. s-cis-trans Isomerism

Double Bonds Connected by a Single Bond. Conjugation of double bonds through a single bond leads to a considerable increase in the double-bondedness of the latter and an increase in the energy barrier of rotation about the single bond amounting to 10-20 kcal/mole. As a result, cis-trans isomers relative to the single bond (s-cis-trans isomerism [50]) arise which can be identified by means of various physical methods. Since the s-cis-trans isomers generally differ considerably in their polar properties, to study their structure and interconversions the dipole moment method is frequently used.

In the simplest case, when the s-trans form is centrosymmetrical, a measured dipole moment different from zero shows the existence of an equilibrium with the s-cis form. Then the position of the equilibrium of the configurations can be evaluated according to (III.21) when the expected moment of the pure s-cis isomer has been calculated.

Thus, on the basis of the results of microwave measurements, Beaudet [51] determined the effective dipole moment of 1,1,4,4-tetrafluorobutadiene (0.4 D) and found that the content of the s-cis form (I) was only 4%, 96% being s-trans isomer (II).

In a number of cases, on the basis of simple electrostatic considerations it is not difficult to decide which of the isomers is present in the equilibrium mixture in greater amounts. If the molecule has several centers of positive and negative charges, the most stable configuration will be that in which the charges of the same sign are as far away from one another as possible and those of different signs are adjacent. This conclusion can be substantiated by several other examples, besides the case of tetrafluorobutadiene already considered.

Thus, of the two planar configurations of m-nitrobenzaldehyde the s-trans structure (III), in which the nitro group and the car-

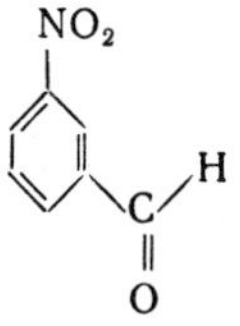

III   IV

$\mu_{calc} = 1.81; \mu_{exp} = 3.22 \ D$ [52] $\mu_{calc} = 5.21 \ D$

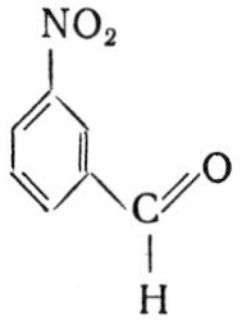

bonyl oxygen, which bear negative charges, are located at a greater distance than in the cis form (IV), predominates (70%).

Conversely, in the case of o-methoxyacetophenone

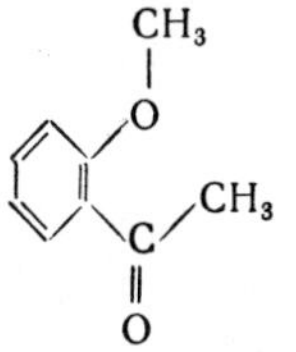

V   VI

$\mu_{calc} = 3.98; \mu_{exp} = 3.39 D$ [53] $\mu_{calc} = 3.43 \ D$

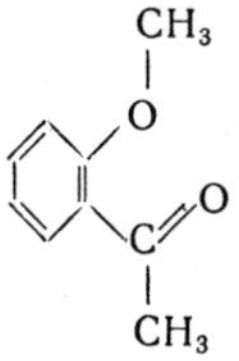

the s-cis form (VI), in which the positively charged ether oxygen is close to the carbonyl group, is produced [53].

In the general case, in order to predict the energetic suitability of one of the configurations, we must calculate the energy of systems of point charges obtained as the result of quantum-mechanical calculations or evaluate the energy of a system of several dipoles in different orientations [54, 55].

Great attention has been devoted to the study of the s-cis-trans isomerism of heterocyclic ketones and aldehydes and their anils. For example, for 5-bromofurfural it is possible to decide between the s-cis configuration (VII) and the s-trans configuration (VIII):*

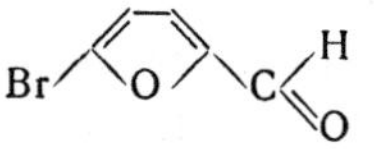

VII   VIII

s-cis   s-trans

$\mu_{calc} = 3.43; \mu_{exp} = 3.37 \ D$ $\mu_{calc} = 1.46 \ D$

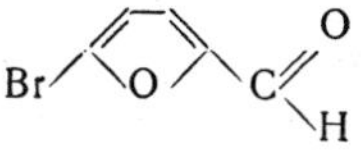

---

* Here and in Table 18 the names indicate the relative positions of the carbonyl oxygen and the ring heteroatom.

TABLE 18. Dipole Moments of Heterocyclic Ketones and
Aldehydes and Their Anils

No.	Compound	μ_{calc}, D		μ_{exp}, D	Literature
		s-cis	s-trans		
1	Furfural	3.54	2.82	3.56	[57]
2	5-Chlorofurfural	3.48	1.35	3.37	[57]
3	5- Bromofurfural	3.43	1.74	3.37	[57]
4	5-Iodofurfural	3.44	2.24	3.29	[57]
5	5-Nitrofurfural	4.61	1.42	3.46	[57]
6	2-Thiophenealdehyde	3.39	2.82	3.48	[58]
7	2-Acetylthiophene	3.40	2.62	3.36	[58]
8	2-Pyrrolealdehyde	1.94	3.84	1.88	[59]
9	2-Acetylpyrrole	1.21	4.13	1.52	[59]
10	2-Pyridinealdehyde	5.05	3.51	3.35	[60]
11	2-Acetylpyridine	5.15	2.95	2.84	[60]
12	2-Furylideneaniline	2.28	0.95	2.01	[61]
13	2-Thienylideneaniline	2.17	1.04	2.21	[61]
14	2-Pyrrolylideneaniline	0.58	3.29	0.87	[61]
15	2-Pyridylideneaniline	3.67	1.14	1.43	[61]

By calculating the moments of (VII) and (VIII) as the sums
of the vectors of the moment of furan (0.71 D), which is in the direction of the oxygen atom along the bisector of the internal angle,
the moment of the C —Br bond, determined from the dipole moment
of α-bromofuran [56], and the moment of the aldehyde group
(Table 10), the C —C axis of which makes an angle of 125° with
the bonds of the furan ring, we find the values μ_{VII} and μ_{VIII} . A
comparison of these figures with the experimental result is in
favor of the s-cis configuration (VII) [57].

Table 18 gives the experimental values of the dipole moments of a series of heterocyclic ketones, aldehydes, and anils
with the values calculated for the s-cis and s-trans configurations.
It follows from the figures given that all the derivatives of five-
membered heterocycles exist in solution in the form of the s-cis
configurations. Conversely, derivatives of the six-membered
heterocycle pyridine (Nos. 10, 11, and 15) are present predominantly in the s-trans form. In the case of 5-nitrofurfural, the two
configurations are approximately equiprobable.

These conclusions are in very good agreement with the results obtained by means of IR spectra [62], PMR spectra [63, 64], and the Kerr effect [65, 66].

The necessity of using data based on other methods of investigation to establish the configuration of compounds of the type considered may be illustrated with α-diketones as an example.

The dipole moment of biacetyl is 1.08 D [65]. This value can correspond either to an equilibrium of the planar s-cis (IX) and s-trans (X) configurations

$$\underset{\substack{\text{IX}\\ \mu_{\text{calc}}\,=4.73\ D}}{\overset{\substack{H_3C\\ O}}{\underset{O}{\diagup}}{C-C}{\overset{CH_3}{\underset{O}{\diagup}}}} \rightleftarrows \underset{\substack{\text{X}\\ \mu_{\text{calc}}\,=0}}{\overset{\substack{O\\ H_3C}}{\underset{}{\diagup}}{C-C}{\overset{CH_3}{\underset{O}{\diagup}}}}$$

or to a rigid nonplanar structure (XI, R = CH$_3$) with the angle φ being between 0° and 180°. The magnitude of the angle φ can be calculated from the formula

$$\mu^2_{\text{exp}} = 2\mu_{\text{COR}} \sin^2 \theta\,(1 + \cos \varphi) \tag{IV.3}$$

where μ_{COR} is the group moment and θ is the angle of inclination of the vector of the group moment to the C$-$C axis. Calculation for biacetyl leads to $\varphi \simeq 160°$, i.e., to a slightly skew s-trans configuration. The choice between this possibility and an equilibrium of the planar configurations (IX)$\rightleftarrows$(X) was made in favor of (XI) since negative values of the molar Kerr constants have been found for α-diketones, while positive values have been calculated for the planar configurations. The greater the volume of the group R, the greater is the deviation from the planar configuration, i.e., the closer is φ to 90°. This is confirmed by the data given below [65]:

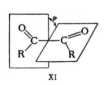

XI

R . . . . $\varphi°$
H   0
CH$_3$ 160
2-Furyl 118.5
C$_6$H$_5$ 97
2-Pyridyl . . . . 81.5 [60]

The configurations of compounds with polyene fragments are the most complex variants of s-cis-trans isomerism. In a number of cases, for example the heterocyclic chalcones [67, 68], the conformations of the molecules in solution can be estimated from the dipole moments.

Double Bond or Aromatic Nucleus Adjacent to a Heteroatom. In all the examples considered above, geometrical isomerism was due to different orientations of fragments of the molecule containing double bonds relative to a single bond. If a single bond is included between even one atom X with an unshared electron pair which is capable of participating in conjugation with the π-electrons of a double bond or an aromatic nucleus, yet another type of s-cis-trans isomerism is possible. For example, for X = O, S, Se, NR':

XII s-trans XIII s-cis

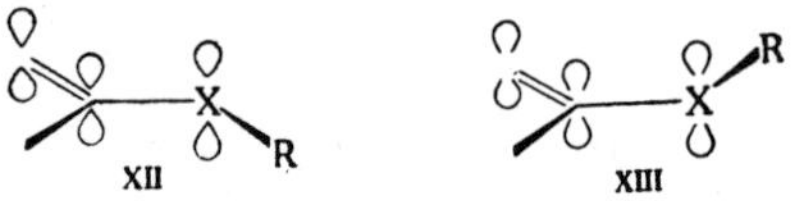

Yet another cause of isomerism is the appearance of an energy barrier to rotation round a single bond, leading to the situation that of all the possible rotational isomers the most stable are (XII) and (XIII), which satisfy the condition of maximum overlapping of the p-electron cloud of the heteroatom with the π-cloud of the double bonds.

If structures of types (XII) and (XIII) have the same spatial energies, because of the identities of their π-electron configurations we may expect that they will be present in solutions or in the gas phase in equal amounts. This assumption is confirmed in the case of the p-dimethoxy- and p-dimethylthiobenzenes.

$\mu_{calc} = 0$ $\mu_{calc} = 2.40\ D$ $\mu_{calc} = 0$ $\mu_{calc} = 2.48\ D$

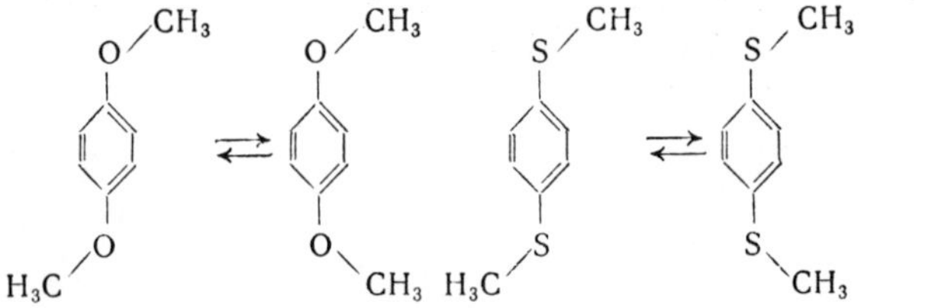

The experimental values of the effective dipole moments of these compounds (benzene 30°C) are 1.71 D for p-dimethoxybenzene and 1.81 D for p-dimethylthiobenzene [69] which, according to (III.21) correspond to contents of the cis form of 51 and 53%, respectively.

TABLE 19. Dipole Moments of o-Substituted
Anisoles and Thioanisoles

| Compound | | μ_{calc}, D | | μ_{exp}, D | Content of |
X	Y	(XIV), s-cis	(XV), s-trans		the s-trans form, %
O	F	1.14	2.50	2.31 [70]	82
O	Cl	1.20	2.60	2.50 [70]	90
O	Br	1.19	2.58	2.47 [70]	90
O	CH$_3$	1.65	1.04	1.00 [71]	100
O	NO$_2$	2.77	4.93	4.83 [72]	94
O	NH$_2$	2.70	1.32	1.55 [73]	88
S	Cl	1.23	2.88	2.56 [74]	74
S	Br	1.21	2.85	2.53 [74]	75

The introduction of a substituent Y into the ortho position of
the benzene nucleus leads to considerable steric hindrance for the
s-cis structure (XIV), which becomes energetically less favorable
than the s-trans form (XV).

This leads to a displacement of the equilibrium of the config
urations in the direction of the s-trans form (XV). The data of
Table 19 confirm the deduction made.

The peri hydrogen atoms of the naphthalene and anthracene
nuclei also create considerable steric hindrance for the planar
configurations of the methoxy and methylthio groups. Thus, it is
obvious that in the case of 1,4-dimethoxynaphthalene the cis configuration must predominate, and for 1,5-dimethoxynaphthalene
and 1,5-dimethoxyanthracene the trans configuration. A comparison of the dipole moments of these compounds with the values calculated for the s-cis and s-trans configurations of p-dimethoxybenzene (see above) agrees well with the hypothesis put forward
[75-77].

If that part of the molecule to which the methoxy group is attached has an asymmetrical distribution of electric charges, the configuration with the minimum energy of electrostatic repulsion will be the most favorable. This factor is less important in the selection of a molecule of stable configuration than steric hindrance. However, in the absence of the latter it may prove decisive. In actual fact, as follows from the value of the effective dipole moment of 2-methoxypyridine and the values calculated for the s-cis (XVI) and s-trans (XVII) conformations

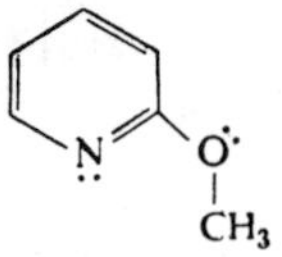

XVI

XVII

$\mu_{calc} = 1.3; \quad \mu_{exp} = 1.06\ D \qquad \mu_{calc} = 3.3\ D$

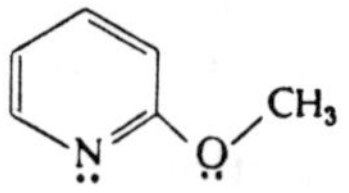

structure (XV), in which the dipoles not participating in the π-bonding of the hybrid electron pairs of the nitrogen and the oxygen are located in an energetically more favorable position, is predominant [73].

Configuration of Complex Functional Groups. Groupings of the type of (XII) and (XIII) are the component parts of many functional groups, for example, −COOR, −CONHR, etc. By analyzing the values of the dipole moments of compounds including these groups it is possible to determine their structure.

For the ester group we may assume either the s-cis structure (XVIII) or the s-trans structure (XIX).

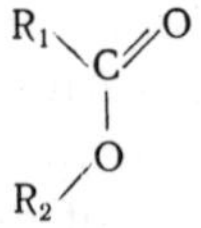

XVIII

$\mu_{calc} = 3.5\ D$

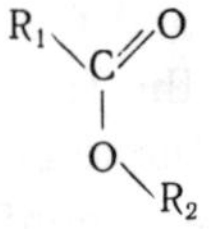

XIX

$\mu_{calc} = 1.50\ D$

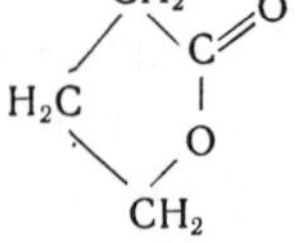

XX

$\mu_{exp} = 3.82\ D\ [78]$

The experimental values of the dipole moments of esters of aliphatic carboxylic acids vary according to the nature of R_1 and R_2 between 1.45 and 2.0 D [3] and depend only slightly on the temperature. This latter fact shows the rigidity of the configuration of an ester. A comparison of the experimental values of the dipole moments with the figures calculated from the moments of the individual bonds for each configuration [75] is in favor of the trans structure (XIX). An additional confirmation is a comparison with the dipole moment of γ-butyrolactone (XX), which, for steric reasons, can exist only in the cis form of type (XVIII). However, for macrocyclic lactones having from ten to sixteen links in the ring and possessing high conformational lability, a trans configuration of the lactone group of type (XIX) exists, as follows from the values of their dipole moments, which are only 1.8-2.0 D [79]. The figures given agree with the energetic preferableness of the trans structure. It has been shown that the trans form of methyl formate (R_1 = H, R_2 = CH_3) is 2.7 kcal/mole more favorable than the cis form [80, 81]. It is obvious that this magnitude will rise with an increase in the dimensions of the R_1 and R_2 groups.

The planar s-trans structure is also characteristic for the monomers of aliphatic carboxylic acids [82] and their sulfur analogs. This is shown by a comparison of the experimental values of the dipole moments of thioacetic acid (R = CH_3) and thiostearic acid (R = $C_{17}H_{35}$) with the moments calculated for the s-trans (XXI) and s-cis (XXII) configurations [83]:

$$\mu_{calc} = 1.64; \ \mu_{exp} = 1.9 \, D \qquad \mu_{calc} = 2.95 \, D$$

Similar calculations carried out for esters of the thiol [83] and thione [84] forms of carbothioic acids lead to the conclusion that they exist in the trans configurations (XXIII) and (XXIV), respectively.

The problem of establishing the configuration of the imide group is somewhat more complex. Even for N-methyldiacetamide it is possible to envisage three planar structures

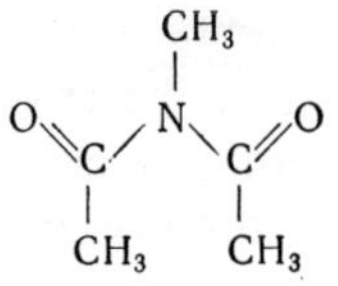

XXV

trans-trans

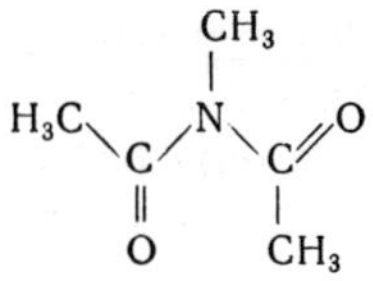

XXVI

cis-trans

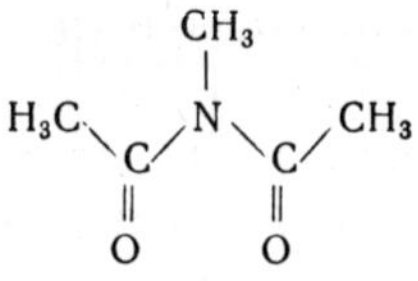

XXVII

cis-cis

Rough calculations of the values of the dipole moments show that structure (XXVII) must be the most polar ($\mu_{cal} > 6$ D), but there is no appreciable difference between the values of the moments of (XXV) and (XXVI), taking the approximate nature of the calculations into account: $\mu_{XXV} \simeq \mu_{XXVI} \simeq 3$ D. A comparison of these figures with the experimental values of the dipole moment of N-methyldiacetamide, which varies according to the solvent between 3.0 and 3.2 D, enables structure (XXVII) to be eliminated but does not make it possible to decide between (XXV) and (XXVI). Much light is thrown on the problem by a comparison with the dipole moments of the model compounds N-methylglutarimide (XXVIII) and N-acetylpiperidone (XXIX):

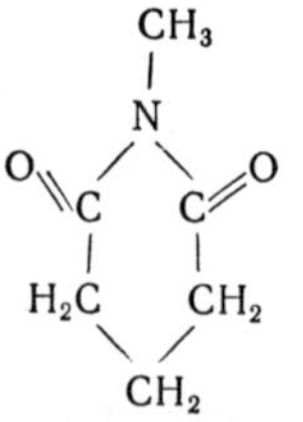

XXVIII

$\mu = 2.70\,D$

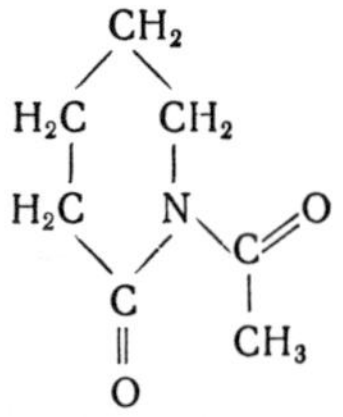

XXIX

$\mu = 3.22\,D$

Compounds (XXVIII) and (XXIX) exist in the fixed trans-trans and cis-trans configurations, respectively (the low value of the dipole moment of (XXIX) shows that the cis-cis structure is impossible). Since the experimental value of the dipole moment of N-methyldiacetamide is closer to that of (XXIX) than to that of (XXVIII), Lee and Kumler [85] concluded that it had the cis-trans configuration (XV). This is confirmed by the results of IR spectra

4. Nonplanar Conformations of Aryl

Nuclei in Uncondensed Aromatic Systems

Conformations of Biphenyls and Their Analogs. A special case of isomerism with respect to an ordinary bond conjugated with double bonds is the rotational isomerism of the biphenyls. The minimum configurational electronic energy is achieved when the angle between the planes of the aryl nuclei $\varphi = 0°$. However, in view of the steric interaction between the o,o'-hydrogen atoms or the o,o'-substituents, which is a minimum at $\varphi = 90°$, the most stable configurations of the biphenyls prove to be intermediate between the structures with the coplanar and the perpendicular orientations of the nuclei.

It is possible to determine the angle φ for the stable conformations of the biphenyls from the dipole moments of o,o' or m,m' disubstituted biphenyls. The method of calculation is shown by the following example:

$$
\begin{array}{lll}
& Cl\ (1) & Cl\ (2) \\
m_x = 0 & m_x = \mu_{Cl} \cos 30° \sin \varphi \\
m_y = \mu_{Cl} \cos 30° & m_y = \mu_{Cl} \cos 30° \cos \varphi & \text{(IV.4)} \\
m_z = \mu_{Cl} \sin 30° & m_z = -\mu_{Cl} \sin 30°
\end{array}
$$

The dipole moment of o,o'-dichlorobiphenyl can be regarded as the vectorial sum of the moments μ_{Cl} contributed by the two chlorophenyl nuclei rotated by an angle φ. The projections of the moments of each of the nuclei on the coordinate axes can be represented in the form of (IV.4). Using equation (III.7), we obtain

$$
\mu^2 = 2\mu_{Cl}^2 \cos^2 30° (1 + \cos \varphi) \tag{IV.5}
$$

From the experimental value $\mu = 1.97$ [86] and $\mu_{Cl} = 1.59$ D (see Table 10), we can calculate $\varphi = 88°$. It is easy to show that formula (IV.5) is also suitable for calculating the dihedral angles φ in the case of similarly m,m'-disubstituted biphenyls. Table 20 gives figures for the conformations of the substituted biphenyls obtained in various investigations.

The values of the angles φ found show that the most stable conformations of the biphenyls are those in which the aryl nuclei

TABLE 20. Dipole Moments and Conformations of Substituted
Biphenyls (Benzene, 25°C)*

No.	Substituent	Dipole moment, D	Calculated angle φ,°	Literature
1	2,2'-Difluoro-	1.88	79	[86]
2	2,2'-Dichloro-	1.97	82(74)	[86]
3	2,2'-Dibromo-	1.90	85(75)	[86]
4	2,2'-Diiodo-	1.71	84(79)	[86]
5	2,2'-Dinitro-	5.10	80	[86]
6	3,3'-Dichloro-	1.80	97	[90]
7	3,3'-Dinitro-	4.17	105	[86]
8	2,2'-Dimethyl-5,5'-dichloro-	2.37	90	[91]
9	2,2'-Dimethyl-5,5'-dinitro-	5.53	90	[91]
10	2-Methyl-5,5'-dinitro-	4.55	104	[91]

* For compounds 1-5 a correction for the induced moments (Chapter V) was introduced into the calculations of the angle φ, decreasing the angle calculated
from formula (IV.5) by about 5-10°. The results of crystallographic determinations of φ [87-89] (accuracy ±5°) are given in brackets. For compounds
6-10, the moments induced in the different nuclei cancel one another out.

are arranged in approximately perpendicular planes. Such a structure is confirmed by results based on a study of the PMR spectra
of the biphenyls [92].

On the basis of the figures of Table 20 we may conclude that
although the deviations from the conformations with mutually perpendicular aryl nuclei are slight, in the case of o,o'-disubstituted
biphenyls the cis arrangement of the polar substituent predominates, and in the case of the m,m'-disubstituted biphenyls the trans
arrangement. This conclusion is in good agreement with the results of X-ray structural studies [88, 93, 94].

Cis configurations with a nonplanar arrangement of the nuclei are also characteristic for 2,2'-disubstituted 9,9'bianthryls
(II) [95]:

X	μ, D	φ, °
F	2.18	85
Cl	2.51	62
Br	2.69	76

II

On the other hand, the shortening of the bond between the condensed nuclei as, for example, in the bixanthylene derivatives (III) and (IV) increases steric interaction between the peri hydrogen atoms and leads to conformations with perpendicular nuclei [96].

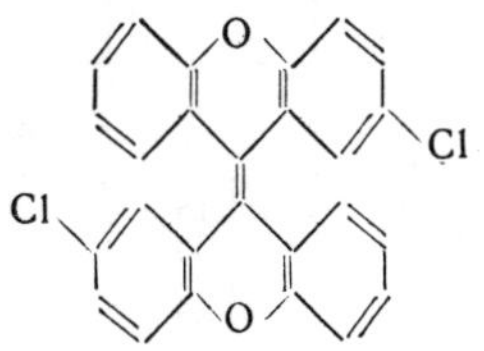

III
$\mu = 2.00\,D$
$\varphi = 89°$

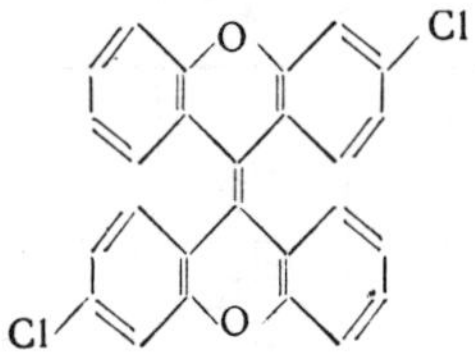

IV
$\mu = 2.07\,D$
$\varphi = 93°$

Conformations of Aryl Nuclei in Compounds of the Type Ar—M—Ar. For compounds of this type [for example (V)] the planar configurations ($\varphi = 0°C$) are sterically unrealizable. Vector calculations carried out for the diphenyl ethers (M = O) [97] and diaryl sulfides (M = S) [98, 99] containing substituents in the o,o'- or m,m'-positions show that those conformations in which one or both aryl nuclei

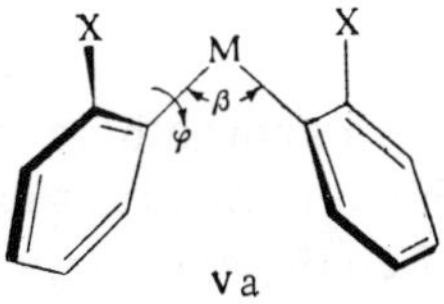

va 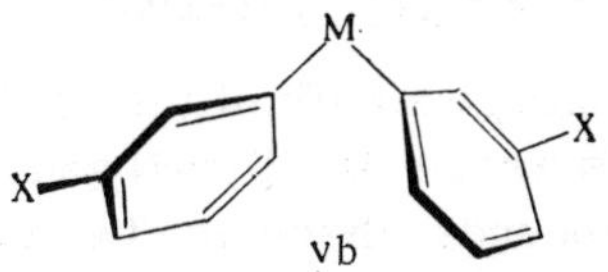   vb

are located at an angle of 90° to the CMC plane are also unrealizable. Consequently, the only remaining possibilities are conformations in which both aryl nuclei deviate from the plane of the CMC bonds. The angle φ which the plane of the aryl nucleus forms with the CMC plane can be determined from the experimental values of the dipole moments of compounds (V) containing one or several substituents in the o,o' or m,m' positions.

Calculation of conformations of type (V) with two similar substituents can be carried out by means of the formula [99]

$$\mu_{exp} = \mu_{Ph_2\,M} - 2\mu_X\left(\cos\alpha\cos\frac{\beta}{2} + \sin\alpha\sin\frac{\beta}{2}\cos\varphi\right) \qquad (IV.6)$$

where μ_X is the group moment of a regular group X and α is the angle formed by the C—X and C—M bonds [60° for (Va) and 120°

TABLE 21. Dipole Moments and Conformations of
Compounds (V)

Compound			$\beta°$, [88]	μ_{Ph_2M}, D	μ_{exp}, D	$\varphi,°$	Litera-ture
type	M	X					
(Va)	O	I	123	1.19	2.72	66	[97]
(Va)	O	NO$_2$	123	1.19	6.64	55	[97]
(Va)	S	CH$_3$	107	1.55	1.18	73	[99]
(Va)	S	Cl	107	1.55	3.33	68	[99]
(Vb)	S	CH$_3$	107	1.55	1.73	86	[99]
(Vb)	S	Cl	107	1.55	1.98	54	[99]

for (Vb)]. Table 21 gives the experimental values of the dipole moments and angles φ calculated for several compounds of type (V).

Nonplanar conformations of the aryl nuclei have been established for diaryl sulfoxides (M = SO) and diaryl sulfones (M = SO$_2$) [98, 99] and also follow from the values of the dipole moments of m,p'-disubstituted diphenylmethanes [100]. These conclusions are confirmed in an analysis of the molar Kerr constants for the compounds Ar−M−Ar* [101–103].

The acoplanarity of the aryl nuclei in azomethine and azo compounds where M = CH=N and N=N have a different nature. In these compounds, there are two types of conjugation (π,π and n,π), the competition between which leads to a departure of the N−aryl nucleus from the plane of the double bond of the bridge group [104, 105]. Calculation of the angle between the planes of an aryl nucleus and the double bond can be carried out on the basis of the dipole moments of o- or m-substituted derivatives of type (VI). However, since conformations (VIa) and (VIb) are almost equiprobable and $\mu^2_{exp} = 0.5\mu^2_{VIa} + 0.5\mu^2_{VIb}$, the effective dipole moment is almost independent of φ.

At the same time, in the azomethines (VII) a simple intramolecular hydrogen bond fixes only one nonplanar conformation of

---

* According to recent refined data, the analysis of the molar Kerr constants may agree both with conformations of type (V) and conformations in which the two aryl nuclei are in mutually perpendicular planes.

type (VIa), while in the conformation of type (VIb) an intramolecular hydrogen bond is sterically incapable of existing. An investigation of the IR spectra of o-hydroxyazomethines shows that practically all their molecules exist in the conformations (VII). Calculation of the dihedral angle φ by a vectorial scheme taking the moments of the individual fragments of the molecule (V) into account
leads to values of φ between 30° and 60° [61], which is in good
agreement with the results of quantum-mechanical calculations
[105].

In the case of azo compounds, in view of the influence of the
effect described above, both aromatic nuclei must deviate from the
plane of the double bond, but the angle of deviation will be less
than for the azomethines, since there is no steric overlapping of
the ortho hydrogen atoms of the aromatic nuclei with the hydrogen
of the methine group preventing coplanarity. In actual fact, calculations and measurements of the dipole moments of trans-2,2'-azo-
pyridine, trans-3,3'-azopyridine, and trans-6,6'-azoquinoline lead
to values of φ of 18-28° [106].

The case of the cis isomers of the azopyridines (VIII) is interesting. Three pairs of conformations are possible for the 2,2'
and 3,3' isomers (the letters denote the positions of the nitrogen
atoms):

		VIIIa	VIIIb	VIIIc
2,2'-azopyridine				
($\mu = 4.04\,D$)		(ab')	(ba')	(aa')
3,3'-azopyridine				
($\mu_{exp} = 2.86\,D$)		(cd')	(dc')	(cc')

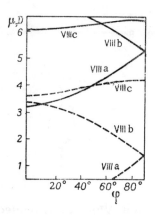

Fig. 21. Calculated dipole moments of the cis isomers of 2,2'-azopyridine (full line) and of 3,3'-azopyridine (broken line) as functions of the angle φ [106].

Each conformation corresponds to its own value of the dipole moment. The dependence of these values on the angle φ is shown in Fig. 21. It is easy to see that the experimental value of the dipole moment for cis-2,2'-azopyridine can be matched only with the conformation (VIIIa), in which the angle $\varphi \simeq 56°$. Similarly, cis-3,3'-azopyridine exists in the conformation (VIIIb) ($\varphi \simeq 40°$). The values of the angles φ obtained agree well with X-ray structural data ($\varphi = 56°$ for cis-azobenzene) [26, 28].

5.   Conformational Analysis of Saturated Cyclic Systems

The determination of the geometrical structure of saturated cyclic systems is one of the most difficult problems of structural organic chemistry, and its solution usually requires the combined use of several physical and chemical methods of investigation. In the present section the role of the dipole moment method in the study of the conformations of small, medium, and some higher cycles is characterized.

If a molecule containing polar bonds the structure of which is to be established can assume one of several conformations possible for it which differ in the magnitudes of the expected dipole moments, an experimental determination of the latter and a comparison with the values calculated theoretically for the different conformations enables us, in a whole series of cases, to make a reliable decision in favor of one of them. In the same way, a value intermediate between two or several calculated values of the dipole moments can be used as an indication of an equilibrium of conformations and be a basis for evaluating the constant and the thermodynamic parameters of the conformational equilibrium.

TABLE 22. Summaries of Data on the Geometry of Some Cyclic
Systems

Initial hydrocarbon or heterocycle	Configuration	Literature
Bicyclo[1,1,1]pentane	–	[107]
Bicyclo[2,1,1]hexane	–	[107]
Bicyclo[2,2,1]heptane	–	[107], Table 15
Cyclopentane	Envelope, half-chair, planar conformation	[108, 109]
Cyclohexane	Chair, boat, skew conformation	[110, 111], Table 13
Cyclohexanone	Chair, boat, planar conformations	[112, 113], Table 14
Cyclohexene	Half-chair	[110]
Endoxocyclohexane	–	[114]
Dibenzobicyclo[3,2,1]-octadiene	–	[115]
Dibenzobicyclo[2,2,2]-octadiene	–	[115]
Bicyclo[2,2,1]heptene	–	[116,117]

Such an approach is obviously feasible only for those compounds of which the molecules contain, as a minimum, two polar
bonds or groups the relative orientation of which changes in the
different conformations. The necessity for the latter condition becomes obvious in a consideration of, for example, molecules of the
type of tetrahydrofuran òr piperidine, the various conformations of
which have the same arrangement of the polar bonds to which identical values of the calculated dipole moments (calculated without
taking induction into account) correspond.

The strain of such rings, leading to small changes in the hybridization of the carbon atoms in the ring, and the dependence of
the induced moments on the orientation of the polar groups lead to
the necessity for these effects to be taken into account more rigorously than in aromatic systems. The most common approach in
this situation is an evaluation of the bond moments and the moments of the individual conformations from the dipole moments of
model compounds.

The principles of calculating the dipole moments expected
for the individual conformations of the cycles and the configurations of the polar substituents have been described in Chapter III.

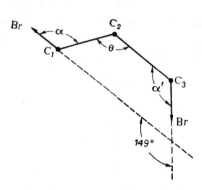

Fig. 22.   Geometry of 1,3-dibromo-
cyclobutane.

Success in the calculations depends very strongly on whether the calculation is based on an adequate geometry of the molecule. Apart from the figures given in Tables 13–15, at the present time information has accumulated on the geometrical characteristics (coordinates of the atoms and the individual bond vectors) of many other cyclic systems. Information on the most important of them is given in Table 22.

Four-Membered Rings

The investigation of the dipole moments of cyclobutane derivatives has given the first clear indications of the nonplanar structure of the cyclobutane ring [118, 119]. By means of dipole moments it has also been possible to calculate fairly accurately the dihedral angle θ which characterizes the degree of deviation of the conformation of cyclobutane from a plane [119–121].

$$\text{(IV.7)}$$

Let us consider the basis of such a calculation with trans-1,3-dibromocyclobutane as an example [121]. The behavior of this compound in solution can be described as an equilibrium inversion (IV.7) leading to two equivalent conformations.

The experimental value of the dipole moment of trans-1,3-dibromocyclobutane is 2.02 D, and that of monobromocyclobutane is 2.03 D [121]. From these figures it is not difficult to calculate from formula (III.5) the angle formed by the vectors of the two dipoles of the C–Br bonds, which is 149° (Fig. 22). Taking into account, also, the fact that the sum of the angles of the quadrilateral is 360° and denoting the angles between the C_1–Br and C_1–C_2 bonds and between the C_2–C_3 and C_3–Br bonds by α and α', respectively, we obtain

$$(180° - \alpha) + \theta + \alpha' + (180° - 149°) = 360°;$$
$$\theta = 149° + \alpha - \alpha'$$

If the extremely reasonable assumption of the equality of α and α' is made, the dihedral angle θ is 149°. This value is in good agreement with the results obtained by other methods of study: for example, with the value θ = 150.5° calculated from an analysis of the microwave spectrum of bromocyclobutane [122].

A nonplanar structure is also characteristic for a hetero-cyclic analog of cyclobutane — thietane. As Arbuzov et al. have shown [123], the experimental value of the dipole moment of 3-chlorothietane does not agree with that calculated for a planar structure (μ_{calc} = 1.74 D) and corresponds to a conformation with an equatorial arrangement of the chlorine atom I with the dihedral angle θ of 143°.

I

μ_{calc} = 1.03; μ_{exp} = 1.01 D

II

μ_{calc} = 2.58 D

Five-Membered Rings

It is known [124, 125] that the cyclopentane molecule is not planar and cyclopentane itself exists in the vapor phase and in solution in the form of a number of energetically approximately equivalent folded configurations existing in a dynamic equilibrium with one another (pseudo-rotation). The most stable forms are the envelope (I) and the half–chair (II).

The introduction of substituents into the molecule of cyclopentane increases the energy barrier between these two forms and, depending on the features of the structure of the particular derivative, stabilizes the envelope or the half-chair conformation. The decision between the alternative conformations for disubstituted cyclopentanes can be made on the basis of their dipole moments in a number of cases. Thus, the dipole moment of 16α–bromo-5α–androstan-17–one measured in benzene is 3.85 D and agrees with the value (3.82 D) calculated for an envelope-like configuration of the cyclopentane ring (I) [108], while that calculated for the half-chair conformation leads to a value of the moments of ~2 D.

It is interesting to note that when the α-halocyclopentanone ring is not condensed in the polycyclic system of a steroid as in (I) the half-chair conformation (II) becomes more favorable for it. This conclusion follows from an evaluation of the angle between the C=O and C—Cl bonds made on the basis of the value of the dipole moment of the compound [126]. The calculations show that this angle must be about 77°, a figure in good agreement with that predicted theoretically for the half-chair conformation but differing considerably from the 94 and 60° expected for the envelope conformation and the plane model, respectively.

In the calculation of the dipole moments of bromocyclopentanones the following values of the moments of the polar bonds are used: $\mu_{C-Br} = 2.10$ D and $\mu_{C=O} = 2.91$ D. The latter value was calculated from the dipole moment of α-chlorocamphor, which contains a cyclopentanone ring. The calculations are facilitated by the existence of a detailed review of the unit vectors of tetragonally and trigonally hybridized substituents on the carbon atoms of cyclopentane in various conformations [108].

Interesting results have been obtained [109] in a study of the conformational equilibrium of the trans-1,2-dihalocyclopentanes. The energy barrier between the envelope and the half-chair conformations here is very small and the conversion between the diaxial and diequatorial isomers leads to a complex equilibrium:

(IV.8)

The values of the dipole moments calculated for the conformations (IIIaa) and (IVaa) are ~0.9 D, and those for the diequatorial forms (IIIee) and (IVee) are 3.1 D. The experimental values for
various dihalo derivatives are between 1.3 and 1.7 D. In accordance with (III.21), these results enable us to state that the content
of the diequatorial conformers in the solution does not exceed 10-
30%. A study of the temperature dependence of the dipole moments
of the trans-1,2-dihalocyclopentanes leads to an extremely low
value of the free energy of the equilibrium aa $\rightleftharpoons$ ee of approximately −1.0 kcal/mole.

Of the saturated heterocyclic compounds with five-membered
rings, the 1,3-dioxolanes have been most studied with respect to
their dipole moments. The replacement of two methylene groups
by oxygen atoms leads to the formation of a pseudo-rotation barrier. In accordance with this, Arbuzov and Yuldasheva [127] have
shown that the closest agreement between the experimental and
calculated values of the dipole moments of 1,3-dioxolane and its
halomethyl derivatives is achieved in the envelope conformation
with a dihedral angle with respect to the bisectrix of the OCO bonds
of 145°. At the same time, the half-chair conformation is more
probable for the 2-alkoxy-1,3-dioxolanes.

Six-Membered Rings

The conformations of six-membered saturated rings have
been studied in considerably more detail than the conformations of
any other saturated cyclic systems. This is due in large measure
to the fact that the very principles of conformational analysis
arose and were formulated as a result of the study of the molecular
geometry of the six-membered alicyclic system of cyclohexane.

The use of the dipole moment method for the study of the
conformations of six-membered rings has been most fruitful in determining the conformations of di- and polysubstituted cyclohexanes, substituted cyclohexanones, including steroids containing
cyclohexanone rings, and heterocyclic analogs of cyclohexane in
which one or more methylene groups have been replaced by heteroatomic groupings.

Cyclohexane Derivatives. Monosubstituted cyclo-
hexanes exist in solution in the form of an equilibrium of two chair-
like conformations separated by a comparatively low energy barrier.

$$(IV.9)$$

In view of this, the experimentally determined value of the dipole moment of a monosubstituted cyclohexane is a value averaged (III.21) for the axial (I) and equatorial (II) isomers. If, however, a voluminous nonpolar substituent R, for example, the tert-butyl group, for which the axial configuration R(II) is energetically extremely unfavorable, is introduced into one of the positions of the cyclohexane ring, the equilibrium (IV.9) is shifted almost completely to the left-hand side and the system becomes conformationally homogeneous [125, 128]. In this case, when the 1,4-substituents R and X are present in the cis position with respect to one another, the axial configuration of the polar group X (I) is stabilized, and in the case of the trans structure the equatorial configuration of X is brought about.

Measurements of the dipole moments of compounds of the cyclohexane series with this type of structure have recently led to the conclusion that one and the same polar substituent possesses different group moments in the axial and the equatorial positions [129-131]. The numerical values of these differences are small, but they fall outside the limits of the possible errors of measurement and are completely regular for a series of compounds (Table 23).

The cause of the differences found in the values of the dipole moments of the axial and equatorial conformers is the dissimilar magnitudes of the vectors of the moments induced by the polar substituent in the hydrocarbon fragment of the molecule (see Chapter V, Section 1).

Induction also exerts a fundamental influence on the magnitude of the dipole moments of polysubstituted cyclohexanes, among which the halogen derivatives have been studied in greatest detail. Havinga et al. [131, 132] have shown that the dipole moments of the conformationally fixed trans-aa-1,2-dihalocyclohexanes (III) differ substantially from the values calculated without taking induction into account. The dipole moments found for compounds

TABLE 23. Dipole Moments of Compounds of the
Cyclohexane Series with a Polar Substituent in the Axial or
the Equatorial Position (Benzene, 25°C)

Compound	Configuration of the polar substituent	μ, D	Literature
trans-1-Bromo-4-tert-butylcyclohexane	e	2.25	[131]
cis-1-Bromo-4-tert-butylcyclohexane	a	2.15	[131]
3β-Chloro-5α-cholestane	e	2.31	[131]
3α-Chloro-5α-cholestane	a	2.06	[131]
trans-4-tert-Butylcyclohexanol	e	1.81*	[130]
cis-4-tert-Butylcyclohexanol	a	1.73*	[130]
trans-4-Hexylcyclohexanol	e	1.71	[130]
cis-4-Hexylcyclohexanol	a	1.63	[130]
cis,cis-3,5-Dimethylcyclohexanol	e	1.73	[130]
cis,trans-3,5-Dimethylcyclohexanol	a	1.44	[130]

* The measurements were performed in the gas phase.

(III) agree well with the values (1.1-1.4 D) obtained for a series of
trans-2,3- and trans-5,6-dihalocholestanes [131] and trans-2,3-
dihalodecalins [133], in which the fixation of the diaxial form is
effected by the formation of an additional cyclohexane ring.

$X_1 = X_2 = Cl$ $\mu = 1.20\ D$ $\mu = 3.33\ D$
$X_1 = Cl,\ X_2 = Br$ $\mu = 1.21\ D$
$X_1 = X_2 = Br$ $\mu = 1.23\ D$

$\mu = 3.42\ D$

Taking the value of the moment of the equatorial C−Br bond
as 2.25 D, in agreement with the data of Table 23, it is possible to
calculate a dipole moment of 3.63 D for a trans-diequatorial iso-

mer of (III) — compound (IV). This value exceeds the experimental value [132] by 0.3 D, which is also explained [131, 132] as the consequence of the mutual induction of the two dipoles of the C—Br bonds. The dipole moment of trans-ee-1,2-dichlorocyclohexane can be taken as equal to the figure obtained for $2\alpha,3\beta$-ee-dichloro-5α-cholestane (V) [131], while the value of the moment of the 1,2-diequatorial chlorobromocyclohexane corresponds to the value of 3.44 D for the corresponding chlorobromocholestane [131]. By making use of these figures it is not difficult to calculate the content of the diaxial (VI) and diequatorial (VII) conformers of the trans-1,2-dihalocyclohexanes in the equilibrium (IV.10). The experimental values given below were obtained for benzene solutions of the compounds [3].

(IV.10)

	μ, D	Ratio of the aa and ee forms
$X_1 = X_2 = Cl$	2.67	44 : 56
$X_1 = Cl$, $X_2 = Br$	2.48	55 : 45
$X_1 = X_2 = Br$	2.16	64 : 36

The results obtained show a reduction in the amount of the diequatorial conformer with an increase in the volume of the substituents. This conclusion is easily explained from the point of view of the ratio of the distances by which the substituents in conformations (VI) and (VII) are separated. Although the equatorial positions for such substituents as chlorine and bromine atoms are approximately 0.5 kcal/mole more favorable than the axial positions [125], the repulsion between X_1 and X_2 in conformation (VII) is considerably greater than in (VI), and increases with an increase in the volume of the substituents. To evaluate the accuracy of the results given, we may state that a determination of the content of the diaxial and diequatorial isomers of trans-1,2-dibromo-cyclohexane by means of IR spectroscopy under identical conditions led to a ratio of 58 : 42 [134].

Dipole moment data have played an important part in determinations of the conformation of polyhalo-substituted cyclohexanes. The presence of a large number of polar groups in the

TABLE 24. Dipole Moments of the Isomeric
Hexachlorocyclohexanes (in D)

Position of the chlorine atoms	Calculated with allowance for induction per reference			Without induc- tion *	μ_{exp} [3]
	[137]	[138, 139]	[140]		
1a2e3e4e5e6a (α)	2.20	2.25	2.20	3.26	2.16 *
1e2e3e4e5e6e (β)	0	0	0	0	0
1a2e3e4e5a6a (γ)	2.84	2.93	3.12	4.62	2.84
1a2e3e4e5e6e (δ)	2.20	2.25	2.20	3.26	2.24
1a2e3e4a5e6e (ε)	0	0	0	0	0

* Calculated without taking induction into account.

molecule makes it particularly necessary to take the induced moments into account. While in the case of the dihalo derivatives the
total influence of induction is expressed by differences amounting
to 0.3-0.5 D between the values calculated without taking it into account and the experimental values of the dipole moments, for tetra-
and hexahalocyclohexanes these discrepancies may exceed 1.5-2 D
[135, 136]. At the same time, in taking induction into account by
using the scheme given below (Chapter V, Section 1), it is possible
to achieve good agreement between the calculated and experimental values of the dipole moments of the isomers of hexachlorocy-
clohexane [137].

It is made easier to take induction into account if the induced
moments are included in the calculation implicitly by assigning to
the moments of the individual polar bonds different values according to the axial or equatorial configuration of the bond and the nature of its environment [138, 139]. In pyridine, a similar approach
has been made by Gey [140], who calculated the moments of all the
types realized in the molecules of the polychloro- and polybromo-
cyclohexanes starting from the values of the moments of the seven
mono- and dihalocyclohexanes. The agreement of the calculated
and experimental figures for a large series of tetra, hexa-, and
heptahalosubstituted derivatives was very good. For illustration,
Table 24 gives the results of calculations of the dipole moments of
several isomeric hexachlorocyclohexanes [137-140], which are
compared with the moments calculated without allowing for induction and with the experimental figures.

TABLE 25. Composition of the Equilibrium Mixture (IV.11) of Conformers of α-Halocyclohexanones in Various Solvents from (III.21)

X	μ_X, D	μ_{XI}, D	μ, D			Content of compound (XI), %			Literature
			hep-tane	ben-zene	di-oxane	hep-tane	ben-zene	di-oxane	
F	4.35	2.95	3.76	4.09	4.20	48	23	15	[143]
Cl	4.29	3.17	3.45	3.73	3.91	78	54	37	[142, 144, 145]
Br	4.27	3.20	3.37	3.50	3.64	85	76	62	[144, 146]

Cyclohexanone Derivatives. The most detailed information has been obtained from the α-halocyclohexanones which exist in solution in two equilibrium conformational forms:

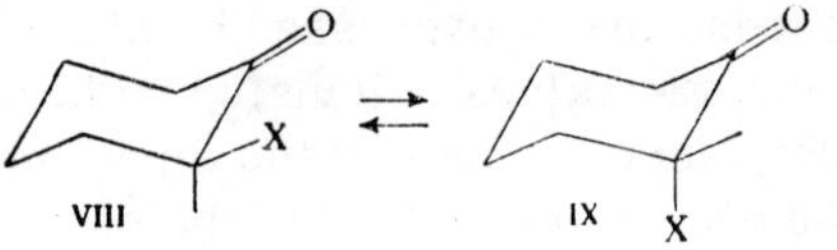

(IV.11)

Starting from the moments of the C$-$X (X = Cl, Br) and C=O bonds for the halocyclohexanes and unsubstituted cyclohexanone and the accurate molecular geometry of the latter (Table 14), and taking the mutual induction of the dipoles into account, particularly in the equatorial conformer (VIII), Allinger et al. [125, 141] calculated dipole moments for (VIII) and (IX) of, respectively, 4.28 and 3.07 D. These values are in good agreement with the moments of the epimers (X) and (XI) the conformations of which are fixed by the 4-tert-butyl group (Table 25).

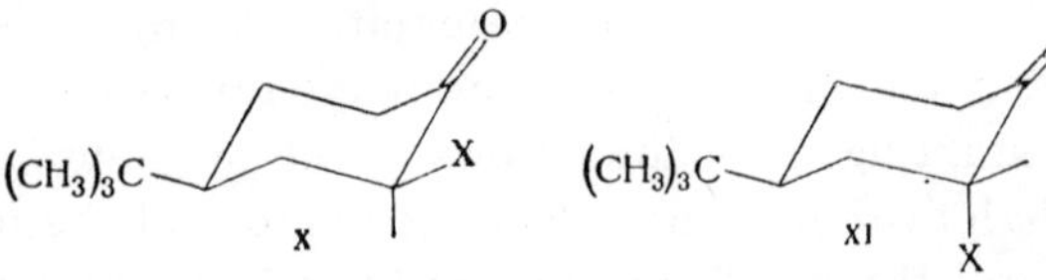

(IV.12)

On the basis of the moments of (X) and (XI) it is possible to determine the composition of the equilibrium mixture of conformers in the equilibrium (IV.11) under various conditions. The results are given in Table 25 and show that the position of the conformational equilibrium depends on the type of solvent and also on

the type of halogen atom. These results are in good agreement
with chemical, spectral, and other studies of the equilibrium (IV.11)
[125, 147].

The determination of the content of each of the conformations
in the equilibrium mixture permits an evaluation of the difference
in free energies ΔG^0 of the conformers at a given temperature

$$\Delta G^0 = - RT \ln K$$

(where K is the equilibrium constant) and a calculation of the conformational energies of the individual groups.

The cyclohexanone system is extremely convenient for such
measurements and calculations. In view of this, the dipole moment
method has been used to study the conformations of many α-halo-
cyclohexanone derivatives [125] and also those of some arylcyclo-
hexanones. Thus, in a study of the conformational equilibrium

$$(IV.13)$$

R = H, $\mu_{calc} = 1.89\,D$, $\mu_{exp} = 1.96\,D$ R = H, $\mu_{calc} = 3.14\,D$
R = CH$_3$, $\mu_{calc} = 1.89\,D$, $\mu_{exp} = 2.56\,D$

it was established [148] that for the compound with R = H the equatorial configuration of the p–chlorophenyl residue (XII) was approximately 1.8 kcal/mole more favorable than the axial configuration (XIII). However, when R = CH$_3$, the two conformations are
approximately equivalent in energy, which may be connected with
the similarity of the effective volumes of the methyl and p–chlorophenyl groups.

The most interesting feature of the cyclohexanone ring, dis-
tinguishing its properties from those of the cyclohexane ring, is
the small difference in the energies of the "chair" and "boat" conformations. Because of its flexibility (pseudo-rotation) the boat
conformation is not fixed and can readily be distorted under the
influence of various structural factors such as the introduction of
substituents or the loading of the molecule with additional cyclic

groupings. This effect is observed in the case of various keto-steroids: derivatives of cholestan-3-one [113, 151] and allobetulone [113], lupan-3-one [113, 149], homoandrostan-3,17-dione [150], and others. The conformations of these compounds have been determined by means of measurements and calculations of the dipole moments carried out for all possible spatial structures.

The first compound of the cyclohexane series studied for which the boat conformation is more favorable (without constraint) than the chair conformation may be taken to be 1,4-cyclohexanedione. This has been found to have a dipole moment in solution [152, 153] and in the gas phase [154] of 1.26-1.44 D. The fact that it has a moment shows the existence of the equilibrium (IV.14).

$$\text{(IV.14)}$$

The amount of conformation (XV) present is easy to calculate when one knows that the angle φ between the two carbonyl groups in (XV) is about 150° and the conformation (XV) is nonpolar. But in this case direct calculation of the type of (III.21) is inapplicable, since the boat (XV) possesses a flexible configuration that is almost equivalent energetically to the enormous number of conformations arising on pseudo-rotation [on the transition from (XV) to (XVI)]. Thus, to calculate the dipole moment of cyclohexanedione we must know the energy function of the pseudo-rotation. Such a function $E(\varphi)$ has been proposed by Allinger et al. [153] on the basis of the general ideas of conformational analysis. The mean dipole moment of cyclohexane calculated by means of this using a formula of the type of (III.19)

$$\bar{\mu}^2 = \frac{\int\limits_{76°}^{180°} 2\mu_{CO}^2 (1 + \cos \varphi) e^{-\frac{\Delta E(\varphi)}{RT}} d\varphi}{\int\limits_{76°}^{180°} e^{-\frac{\Delta E(\varphi)}{RT}} d\varphi}$$

(where 76 and 180° are the limits of the change in φ on pseudo-rotation) agrees fairly well with the experimental figures for the corresponding temperatures. This confirms the correctness of the conformational relations given above for cyclohexanedione.

Six-Membered Saturated Heterocyclic Systems. On the basis of measurements and calculations of Kerr constants, Aroney and LeFevre [155] put forward the hypothesis that a hydrogen atom and a methyl group on the nitrogen of the piperidine ring are predominantly in the axial configurations. The situation can be considered in a simplified fashion as if the un-shared electron pair of the nitrogen atom in piperidine possessed some effective volume which was larger than the effective volume of the hydrogen atom and even than that of the methyl group, since of two substituents on an atom in a ring the equatorial position is occupied by the more voluminous of them. The problem of the "volume" of the unshared pair of the nitrogen caused an extremely lively discussion and was resolved by means of the dipole moment method.

The molecule of N-methylpiperidine may exist in solution in the form of two conformations with the equatorial (XVII) and the axial (XVIII) positions of the methyl group, transition between which takes place both as a consequence of the conversion of the piperidine ring (IV.15) and as a consequence of the inversion of the nitrogen atom (mirror reflection).

$$\text{(IV.15)}$$

A convenient model for establishing the position of this equilibrium is another equilibrium (IV.16) between the conformers of N,N'-dimethylpiperazine [156]:

$$\text{(IV.16)}$$

$$\mu_{calc} = 0 \qquad \mu_{calc} = 1.55\,D \qquad \mu_{calc} = 0$$

The dipole moments of the conformations (XIX) and (XXI) are zero because of symmetry, and the moment of (XX), calculated as the vectorial sum of the dipole moments of two suitably oriented molecules of N-methylpiperidine (μ = 0.95 D) is 1.55 D. The experimental value of the dipole moment of N,N'-dimethylpiperazine in benzene at 25°C is 0.50 D; hence, according to (III.21), the content of the polar form (XX) in the equilibrium mixture (IV.16) is 10%. This figure enables us to calculate the free energy of the reaction (XIX)→(XX), and an entropy calculation from the symmetry numbers leads to a difference in enthalpies between the axial and equatorial methyl groups of 1.7 ± 0.4 kcal/mole in favor of the latter. This magnitude agrees with the enthalpy of the axial methyl group in methylcyclohexane. Similar measurements and calculations have been carried out for mono-N-methylpiperazine. They also show the energetic advantageousness of the equatorial position of the hydrogen, but the barrier is considerably lower (0.4 kcal/mole [156].

A confirmation of the preferability of the equatorial configuration of a methyl group on a cyclic nitrogen was obtained by determining the dipole moment of 4-(p-chlorophenyl)-N-methylpiperidine, which was found to be 2.25 D (benzene, 25°C) [156,157]. Since the equatorial configuration of the aryl group is energetically more favorable than the axial configuration (the difference in free energies is ~2 kcal/mole), this compound exists in solution almost exclusively (to the extent of 97%) in the form of the epimeric conformations (XXII) and (XXIII):

XXII

μ_{calc} = 2.33 D

XXIII

μ_{calc} = 0.88 D

A comparison of the calculated and experimental values of the dipole moments for (XXII) and (XXIII) shows that the equilibrium mixture contains 98% of the conformer with the equatorial methyl group.

The equatorial configuration of the N-methyl group has also been established by means of dipole moments for a number of substituted N-methylpiperidones [158. 162], N-methyltropanes [159, 160], and N-methylmorpholine [161]. In the last case, however, the conformation was found to depend fundamentally on the solvent.

Dipole moment data have been widely used to establish the conformation of a whole series of six-membered saturated cyclic systems containing two and more heteroatoms in the ring. It has been shown that 1,4-dioxane ($X_1 = X_2 = 0$) [163, 164], 1,4-dithiane ($X_1 = X_2 = S$) [164, 165], and 1,4-thioxane ($X_1 = 0$, $X_2 = S$) [164] exist in solutions and in the vapor phase almost exclusively in the form of the chair-like conformation (XXIV). The fact that these compounds

have small dipole moments (~0.5 D) the accuracy of the values of which is low because of the absence of information on the atomic polarizations is generally connected with the presence of a small amount of the polar boat conformation (XXV). The same conclusion applies to 1,2-dithiane [170].

In the case of 1,3-dioxane, not two but three conformations are possible.

$$\mu_{calc} = 1.93\,D \qquad \mu_{calc} = 2.50\,D \qquad \mu_{calc} = 1.11\,D$$

Arbuzov [127, 166], on the basis of a comparison of the dipole moments of 1,3-dioxane and its alkyl derivatives found experimentally (1.80-1.94 D) and the values calculated for the conformations (XXVI)-(XXVIII), concluded that these compounds have the chair conformation (XXVI). The same conformation is preserved for a number of derivatives with polar substituents in position 2 [167], while, according to the dipole moments and the PMR spectra, 5α-methoxyethyl-2,5-dimethyl-1,3-dioxane has the unsymmetrical "boat" conformation (XXVIII) [168].

An extremely wide range of derivatives of 1,3-dioxane and of tetrahydro-1,3-oxazine and hexahydropyrimidine containing a

nitro group in the ring has been studied by Urbanskii [167, 169].
He compared the experimental values of the dipole moments with
those calculated for all the possible conformations. The best
agreement was found for chair-like structures. A particularly interesting feature of this investigation is the finding that the nitro
group has a smaller effective volume than the methyl group on the
same ring carbon atom. Kalff and Havinga [171] have applied the
method of conformational homogenization to a study of the conformation of 2-substituted 1,3-dithianes, using as substituent the p-
chlorophenyl group, which assumes an equatorial position in the
ring. They succeeded in showing that the preferred conformation
of the 1,3-dithianes is a chair-like form analogous to (XXVI).

Chair conformations are also preferred for six-membered
rings with three heteroatoms (XXIX): 1,3,5-trioxane (X = O) [166],
1,3,5-trithiane (X = S) [163, 174], the cyclic trimer of methylene-
aniline (X = C_6H_5N) [175], and trimethylene sulfite [166, 172, 173].

XXIX   XXX

The dimer of aminooxydibutylborane (XXX), belonging to
compounds of the BONBON type, in which a cyclic structure arises
through two N → B coordination bonds, also exists in the chair-like
conformation. The dipole moment of this compound, all the bonds
in the molecule of which are highly polar, is zero, which shows the
absence of appreciable amounts of various boat conformations [176].

In all cases where X-ray and electron diffraction methods
have been used to study the conformations of the compounds described above, their results confirm the assignments made on the
basis of a determination of dipole moments.

Higher Rings and Bridge Systems

The conformations of higher cyclic systems have been studied in considerably less detail than the conformations of saturated
four-, five-, and six-membered rings, and the interpretation of
their dipole moments is made very difficult by the flexibility of the
polycyclic system or the presence of several energetically equivalent conformations.

Borsdorf et al. [177] have measured the dipole moments of 2-halocycloheptanones in several solvents and, judging from the absence of any appreciable solvent influence on the dipole moments, have suggested that these α-halo ketones exist in solutions in the form of a single fairly rigid conformation or a stable equilibrium of several conformations.

It has been possible to derive more concrete information from an analysis of the dipole moments of the cyclooctanones [178]. The dipole moment of 5-(p-chlorophenyl)cyclooctanone (3.39 D) may correspond to the presence of a flexible crown conformation (I) for which the calculated values of the dipole moments vary according to the angle of pseudo-rotation from 1.58 to 3.52 D. A rigid structure with an expected dipole moment of 4.17 D is excluded, and so is the possibility of an equilibrium of the conformations (I) ⇌ (II), because the IR spectra show the presence of only one molecular conformation.

I   II   III

The crown conformation is also realized for a heterocyclic analog of cyclooctanone: oxacyclooctan-5-one (III) [178, 179]. However, in this case it is highly distorted in the direction of a decrease in the dihedral angle between the C−O−C and C−CO−C planes, which is favored by the transannular conjugation of the heteroatom with the carbonyl group (see Chapter V, Section 3).

Investigations have also been made of the dipole moments of eight-membered rings with two and more heteroatoms: derivatives of 1,5-diazacyclooctane [169] and of 1,3,5,7-tetroxacane [180]. A conformation of the crown type (I) is regarded as the most likely.

A considerable number of papers have been devoted to the application of dipole moments to the study of the conformations of bridge structures. We have already (see Chapter III) considered the possibility of determining the position of the polar substituent in the molecules of substituted norbornanes (IV). Kumler et al. [181], using the method of calculation described, has established the syn configuration for a norbornane derivative, 7-chloronorcamphorquinone (V), the experimental value of the dipole moment of which is 5.33 D.

IV

V

$\mu_{calc} = 5.7\,D$

VI

$\mu_{calc} = 2.9\,D$

Even better agreement of the experimental and calculated dipole moments for (V) can be obtained if one takes into account the fact that the moment of the C_7-Cl bond in (IV) is somewhat less than that of the C_2-Cl bond because of small differences in the hybridization of the carbon atoms [182]. The latter circumstance, together with a distortion of the valence angles, also leads to the appearance of a small dipole moment in the bridge hydrocarbon itself, which must be taken into account in the calculation. The greatest dipole moment among the saturated bridge hydrocarbons is possessed by bicyclo[1,1,0]butane, the molecule of which is the most strained of all known cyclic systems. The dipole moment of bicyclo]1,1,0]butane determined by the microwave method is 0.67 D [185].

Dipole moments have been used successfully for determining the position of the substituents in the system of a heterocyclic analog of norbornane (3,6-endoxocyclohexane [114]) and in bicyclo-[2,1,1]hexane [181], bicyclo[3,2,1]octanone [183], and in derivatives of tropane [160]. In the last case, proofs were obtained of a substantial flattening of the six-membered piperidine ring fused to the cyclopentane ring. The effect of the distortion of the conformations of the individual rings is generally characteristic for bridge systems. A typical example confirming this statement and illustrating the possibilities of the dipole moment method for establishing the conformations of bridge structures is formed by the results of an investigation of the conformation of the bis(dibromocarbene) adduct of 1,4-cyclohexadiene — 4,4,8,8-tetrabromotricyclo[5,1,0,0^{3,5}]-octane — to which one of the following possible rigid conformations may be assigned (the angles between the dipoles of the two dibromocyclopropane rings φ and the calculated dipole moments [184] are given):

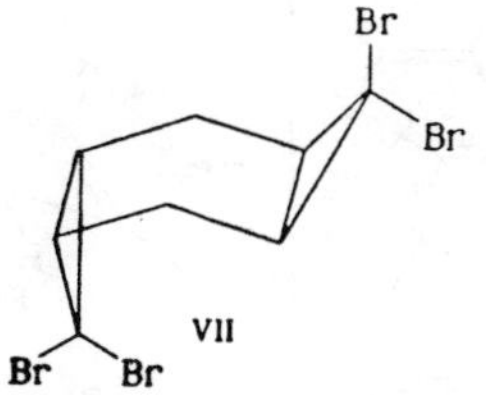

VII

$\varphi = 154°$, $\mu_{calc} = 0.95\ D$

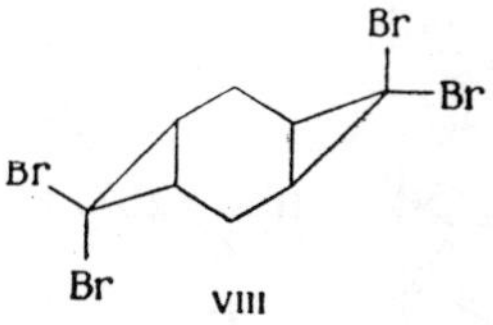

VIII

$\varphi = 180°$, $\mu_{calc} = 0$

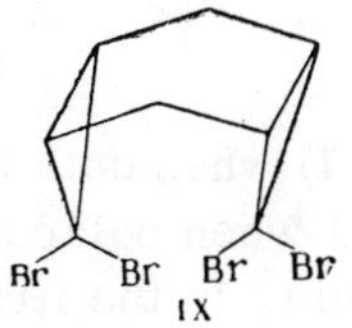

IX

$\varphi = 9°$, $\mu_{calc} = 4.20\ D$

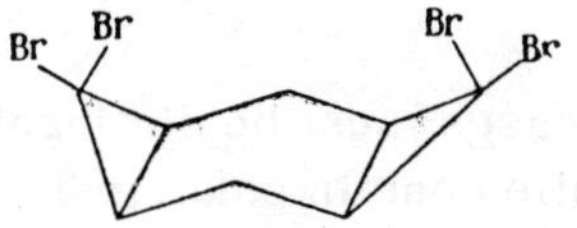

X

$\varphi = 109°$, $\mu_{calc} = 2.76\ D$

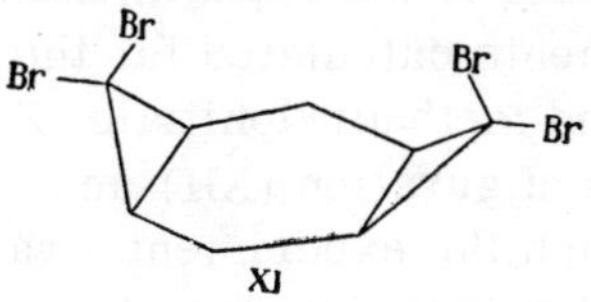

XI

$\varphi = 59°$, $\mu_{calc} = 3.65\ D$

The experimental value of the dipole moment (taking atomic polarization into account) is very close to zero, which corresponds to the trans configuration of the cyclopropane rings relative to the plane of the cyclohexane ring (VIII). The same configuration is characteristic for the unsubstituted hydrocarbon.

A particularly large number of bridge compounds has been obtained by the diene synthesis. The determination of the dipole moments of the adducts of the diene synthesis plays an important part in establishing the principles of the stereochemistry of this reaction, particularly when endo (XII) or exo (XIII) additions take place in dependence on the orientation of the reactants in the transition state, as shown below on the basis of the adducts of cyclopentadiene and its analogs:

Two cases must be distinguished: 1) when both the diene and the dienophile contain polar groups and 2) when polar substituents are present only in the dienophile molecule. In the first case, the dipole moments calculated for the exo and endo structures generally differ so considerably that it is extremely easy to assign the conformations according to the experimental value of the moment. Thus, the dipole moments calculated for the adduct of tetracyclone ($R = C_6H_5$, $Z = CO$) and methacrylonitrile ($X = CH_3$, $Y = CN$) are 1.76 D for the endo configuration (XII) and 5.04 D for the exo (XIII) configuration. Although the experimental value of 2.67 D differs from that calculated for (XII) by almost 1 D, it nevertheless causes no doubt as to the endo structure of the adduct [186].

Similar calculations have been carried out for adducts of various dienophiles with a number of dienes possessing polar groups: furan ($R = H$, $Z = O$) [187], hexachlorocyclopentadiene [117, 186, 188, 190], and 2-nitroanthracene [189], and also for the dimers of trimethylcyclohexadienone [191] and o-quinone [192]. In all these cases the indefiniteness in the selection of the group moments and the geometry of the cyclic system is not reflected in the reliability of the determination of configurations.

Considerably greater care must be used in studying the dipole moments of adducts containing no polar substituents. In addition to a possibly stricter regard to geometry, here we must take into account the moment of the unsaturated hydrocarbon forming the skeleton of the bridge system [193] and, particularly, the mutual induction of the dipoles, which causes a substantial dependence of the bond moments on their orientation in the bridge system

TABLE 26. Determination of the
Configurations of Cyclopentadiene Adducts

Dienophile	μ_{calc}, D		μ_{exp}, D	Config- uration
	exo	endo		
$CH_2{=}CHCl$	1.42	2.33	1.45	exo
$CHCl{=}CHCl$ (cis)	2.85	3.80	2.96	"
$CHCl{=}CCl_2$	2.59	3.18	2.59	"
$CH_2{=}CHCN$	2.96	4.11	3.73	endo
$CHCN{=}CHCN$ (cis)	3.51	4.73	4.97	"

[186-188, 193-196]. This approach is based on an evaluation of
the bond moments from the dipole moments of suitably selected
compounds. Thus, Le Bel [193] has obtained a good agreement of
the experimental dipole moment of cis-5,6-dibromonorbornene
with the figure calculated for the exo structure [(XIII), R = H,
X = Y = Br], using in the calculations the moment of the C−Br
bond found from the dipole moment of the corresponding 2,3-
dibromonorbornane. Arbuzov and Vereshchagin [187, 196] have
used the values of the moments of the C−CN and C−Cl bonds determined from the dipole moments of the corresponding anthracene
adducts to calculate the theoretical moments of various cyano- and
chloro-substituted norbornenes. Table 26 gives some data on the
dipole moments of adducts of cyclopentadiene (R = H, Z = CH_2)
which have enabled their structures to be determined [187, 196,
197].

The data given above do not, of course, exhaust all cases of
the application of the dipole moment method to the determination
of the conformations of saturated or partially unsaturated systems.
We may also refer to studies of the structures of derivatives of
bicyclo[2,1,1]hexane [181], bicyclo[2,2,2]octane [198], benzocyclo-
pentenone [199], tetralin [200, 201], tetrahydroquinoline [200, 203],
4,4-dimethyl-3-ketosteroids [204], and Tröger's base [205].

6. Internal Rotation Relative to Single Bonds

The use of dipole moments has played an important part in
the formation of modern ideas on internal rotation in molecules
and rotational isomerism.

TABLE 27. Dipole Moment of 1,2-Dichloroethane as a Function of the Temperature [3, 209]

Gas		Heptane		Diethyl ether		Gas		Heptane		Diethyl ether	
temperature, °K	μ, D	temperature, °K	μ, D	temperature, °K	μ, D	temperature, °K	μ, D	temperature, °K	μ, D	temperature, °K	μ, D
305.0	1.12	203	1.07	213	1.25	411.6	1.38	–	–	283	1.48
307.3	1.18	223	1.16	223	1.29	419.0	1.40	–	–	293	1.51
337.4	1.21	243	1.24	233	1.32	457.0	1.45	–	–	–	–
341.0	1.24	263	1.31	243	1.35	479.8	1.48	–	–	–	–
353.2	1.25	283	1.36	253	1.39	484.8	1.48	–	–	–	–
376.3	1.32	303	1.41	263	1.43	543.7	1.54	–	–	–	–
384.8	1.34	323	1.42	273	1.45	588	1.57	–	–	–	–

The ideas of free rotation relative to a single bond that were previously widespread were first put in doubt by the results of measurements of dipole moments of molecules with two asymmetric carbon atoms [206]

$$
\begin{array}{cc}
\text{H} \quad C_6H_5 & \quad C_6H_5 \quad \text{H} \\
\diagdown \diagup & \diagdown \diagup \\
\text{X} \quad \text{C} & \text{X} \quad \text{C} \\
| \diagup | & | \diagup | \\
\text{C} \quad \text{X} & \text{C} \quad \text{X} \\
\diagup \diagdown & \diagup \diagdown \\
C_6H_5 \quad \text{H} & C_6H_5 \quad \text{H} \\
\text{I (meso)} & \text{II, } (d, l)
\end{array}
$$

X	μ_{I}, D	μ_{II}, D
Cl [206]	1.27	2.75
OH [207]	2.08	2.67
CN [208]	5.32	4.88

In the case of unhindered rotation around the C−C bond, the meso form (I) and the optically active form (II) would have the same dipole moments. The considerable differences between the moments of these compounds show hindrance to internal rotation, i.e., the existence of a definite potential barrier to rotation.

A further confirmation of this conclusion is given by an investigation of the dipole moments of 1,2-dichloroethane, the magnitudes of which are given in Table 27, and of other 1,2-disubstituted ethanes [82, 209].

The values of the dipole moments given differ substantially from those calculated by formula (III.12) for free rotation. For example, for the gaseous state $\mu_{calc} = 2.47$ D. In addition, the dipole moments of 1,2-dichloroethane and other 1,2-disubstituted ethanes depend considerably on the temperature at which the measurements are made. All this indicates that all the possible isomers produced by internal rotation are not energetically equivalent and shows the necessity for taking into account the potential energy of internal rotation in calculations of the mean dipole moment for each temperature.

The main problem arising in such an approach is finding the type of potential function $V(\varphi)$ that should then be used in calculations using formula (III.19). In early papers on the dipole moments of the 1,2-dihaloethanes [209] attempts were made to determine the function $V(\varphi)$ in the form

$$V = \frac{V_0}{2}(1 - \cos \varphi) \tag{IV.17}$$

where V_0 corresponds to the difference in the energies of the trans and cis rotational isomers and φ is the angle of rotation. Such a function corresponds to the existence of one well-defined potential minimum of the internal rotation (trans form) and enables the temperature dependence of the dipole moments of the 1,2-dihaloethanes to be explained only by introducing the assumption of an extremely broad potential trough. Investigations of the temperature dependence of the dipole moment of 1,1,2,2-tetrachloroethane [82] and, especially, spectroscopic data [209, 210] have shown that the potential energy curve of the 1,2-dihaloethanes and molecules of similar structure contains not one but three minima, the two additional ones of which correspond to skew or gauche forms. Ensembles of molecules of these compounds can be regarded to a good approximation as mixtures of hindered gauche (III, V) and trans (IV) isomers the relative amounts and energies of which can be calculated, together with the magnitude of the potential barrier, from an analysis of the magnitude and temperature dependence of the dipole moment.

As an example, let us consider the calculation of the contents and energy differences of the gauche and trans isomers of 1,2-dichloroethane in carbon tetrachloride at 25°C [211].

It is easy to show that the dipole moment of 1,2-dichloroethane in any conformation defined by the azimuthal angle φ (Fig. 23) can be given in the form

$$\mu\left(\varphi\right) = 2\mu_0 \sin \alpha \cos \frac{\varphi}{2} \qquad (IV.18)$$

where μ_0 is the moment of the chloromethyl group and α is the angle that it makes with the axis of rotation. Assuming that $\mu_0 = 1.72$ D (the moment of methyl chloride in carbon tetrachloride) and $\alpha = 70°$, i.e., the vector of the dipole is directed along the C$-$Cl bond, we obtain

$$\mu\left(\varphi\right) = 3.23 \cos \frac{\varphi}{2} \qquad (IV.19)$$

Making use of relation (IV.19), it is possible to calculate the dipole moments of the trans and gauche rotational isomers of 1,2-dichloroethane. In the case of the trans isomer, $\varphi = 180°$ and $\mu = 0$. From electron-diffraction data, the angle φ for the gauche isomer is 71° [210] and, consequently, its dipole moment $\mu_g = \mu(71°) = 2.63$ D. The experimental value of 1.47 D corresponds to the mean square dipole moment of the equilibrium mixture of rotational isomers

$$1.47^2 = \overline{\mu}^2 = \frac{\mu_t^2 N_t + \mu_g^2 N_g}{N_t + N_g} = \frac{\mu_t^2 + \mu_g^2 \dfrac{N_g}{N_t}}{1 + \dfrac{N_g}{N_t}} \qquad (IV.20)$$

from which it follows that the equilibrium mixture of rotational isomers of 1,2-dichloroethane in carbon tetrachloride at 25°C contains 31% of the polar gauche isomer. The use of the electron diffraction method on 1,2-dichloroethane in the gas phase at 22°C leads to a value of 27% [210].

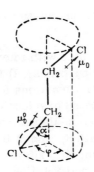

Fig. 23. Scheme for the calculation of the dipole moment of 1,2-dichloroethane.

The difference in the energies of the gauche and trans isomers is calculated by the Boltzmann equation:

$$\frac{N_g}{N_t} = \frac{2f_g}{f_t} \, e^{-\frac{\Delta E}{kT}} \qquad (IV.21)$$

where f_g and f_t are the partition functions of the two conformations and the factor 2 corresponds to the presence in the gauche isomer of two nonidentical forms (III) and (V). The ratio f_g/f_t varies in the temperature range 298-500°K from 0.95 to 0.9 [209]. By using these data it is possible to calculate for carbon tetrachloride $\Delta E = 0.85$ kcal/mole.

In the gaseous state, the trans isomer is still more energetically favored. For this case, using dipole moments and the ratio of the intensities of the lines in the Raman spectrum, Mizushima [209] has calculated $\Delta E = 1.2$ kcal/mole. However, in the liquid state $\Delta E = 0$ and the gauche form already predominates in the mixture of isomers (60% at −10°C). This is explained by the fact that the experimentally determined value of ΔE is in fact composed of two components: the difference in the energies of the isolated molecules of the rotational isomers and the potential energies due to the interaction of each of the isomers with neighboring polar molecules [212]. It is obvious that the reaction field will stabilize the polar gauche conformation. Consequently, on passing from the gaseous state to solutions and to the liquid state, the content of gauche isomer must increase.

In such an approach, all rotational conformations other than the gauche and trans conformations are ignored. In order to take their influence on the dipole moment into account we must know an accurate potential function for internal rotation. For the overwhelming number of molecules with polar groupings, including 1,2-dichloroethane, such functions are unknown. In view of this, calculation of the dipole moment of a molecule with internal rotation must necessarily start from some model, for example the idea of the equilibrium of the gauche and trans rotational isomers, as for 1,2-dichloroethane.

Data on the dipole moments of molecules with internal rotation are required mainly for the following purposes: 1) to decide on the constants in the approximate potential functions for internal rotation the form of which has been postulated previously, 2) to establish the hindrance to internal rotation and determine the nature of the energetically favored conformation, and 3) to check the adequacy or otherwise of the model used for describing the behavior of the ensemble of molecules with internal rotation.

From the structural point of view, of course, the last two aspects are the most important. In an analysis of the dipole moments of molecules capable of possessing internal rotation, they must in the first place be compared with those calculated on the assumption of free rotation. Coincidence of these values, if it is not fortuitous, may be an indication of equal populations of all the rotational levels under the experimental conditions. Conversely, a discrepancy means hindered rotation.

In the simplest cases of rotation about one or several similar single bonds, it is not difficult to determine the conformation the calculated dipole moment of which would coincide with the experimental value. For example, the experimental dipole moment of chloromethyl methyl ether (VI) is 1.90 D in benzene and 1.88 D in carbon tetrachloride (25°C) [213].

$\varphi,°$	μ_{calc}, D
0	0.70
60	1.56
90	2.10
120	2.52
180	2.88

The value calculated for free rotation from equation (III.13) is 2.03 D which, taking certain indeterminacies into account, is fairly close to the experimental value. However, electron diffraction [214, 215] and spectral data [216] for chloromethyl methyl ether show with the definiteness that is characteristic of them that in the gaseous state and in solution its molecules exist in the form

TABLE 28. Dipole Moments of Compounds of the Type AB_4
(Benzene)

Compound		μ_{calc} according to (III.13), D	μ_{exp}, D	Literature	Compound		μ_{calc} according to (III.13), D	μ_{exp}, D	Literature
A	B				A	B			
C	CH_2F	3.36	0.77	[217]	Si	OCH_3	2.06	1.75	[221]
	CH_2Cl	3.52	0.63	[217]		OC_3H_7	2.06	1.66	[220]
	CH_2Br	3.42	0.22	[218]		$OC_{10}H_{21}$	2.06	1.69	[220]
	CH_2I	3.00	0	[218]		SCH_3	2.21	1.41	[221]
	CH_2OCH_3	0	1.91	[219]					
	CH_2NO_2	5.82	2.48 *	[218]	Ge	OCH_3	2.09	1.89	[221]
	OCH_3	2.09	0.83	[220]		SCH_3	2.21	1.68	[221]
	SCH_3	2.21	0.56	[221]					

* In dioxane.

of a single hindered conformation. Calculation of the dipole moments of the various conformations of (VI) differing by the angle of rotation φ lead to a coincidence of the experimental and calculated values for the conformation with $\varphi = 78°$. This figure agrees excellently with the value of 74° found from electron diffraction measurements (VI). At the same time, it is obvious that the conformation with an angle $\varphi = 180°$ assumed on the basis of spectral data is erroneous.

Table 28 compares the experimental values of the dipole moments of the general formula AB_4, where B is a radical the rotation of which about the A−B bond is reflected in the dipole moment of the molecule, with the values calculated on the assumption of free rotation.

It can be seen that for compounds with a central carbon atom rotation relative to the C−B bond is hindered. The observed values of the dipole moments do not depend on the temperature, which shows the absence of an equilibrium of several conformations and is in harmony with the existence in solution of a single fixed structure. Its geometrical form can be determined by finding the conformation the calculated dipole moment of which agrees with the experimental value. In this way the stable conformation (VII) has been found for methyl formate [221].

A closer agreement between the experimental values of the dipole moment and the magnitudes calculated on the assumption of

TABLE 29. Dipole Moments of α,ω-Dibromoalkanes
$Br(CH_2)_nBr$ (Benzene, 25°C) [225]

n	μ_{exp}, D	$\mu_{calc, fr. rot}$, D *	$\mu_{calc, hind}$, D	n	μ_{exp}, D	$\mu_{calc, fr. rot}$, D *	$\mu_{calc, hind}$, D
3	2.066	2.710	2.053	7	2.441	2.766	2.439
4	2.067	2.747	2.081	8	2.480	2.766	2.477
5	2.284	2.760	2.286	9	2.543	2.766	2.542
6	2.328	2.764	2.328	10	2.568	2.766	2.572

* According to formula (III.13) with μ_{C-Br} equal to 1.956 D (the dipole moment of butyl bromide according to the literature [225, 226]).

free rotation is obtained for compounds of the type AB_4 in which the central atom is germanium or silicon; however, in this case as well, internal rotation is hindered [221, 222]. This conclusion is in agreement with the results of electron-diffraction studies [223].

In individual cases the problem of determining the structure of a stable conformation of a molecule with internal rotation of the bonds is facilitated by the fact that in some theoretically predictable conformations conditions are created for intramolecular hydrogen or coordination bonds stabilizing the molecule. As an example we may refer to dialkyltin diacylates, the molecules of which are present in solution in the form of octahedrally coordinated conformations (VIII). This structure is also confirmed by the results of IR spectroscopy [224].

VIII

Particular interest is offered by an analysis of the dipole moments of compounds with long molecular chains and a large number of axes of internal rotation, for example α,ω-substituted alkanes. Compounds of this type with long polymethylene chains

can be used as convenient models facilitating a transition to the
consideration of complex polymeric molecules. Table 29 gives
measurements with a high degree of accuracy (± 0.004 D) of the
dipole moments of polymethylene dibromides. These values are
comparable with those calculated on the assumptions of the free
rotation of all the links $\mu_{calc, fr. rot}$ and with the values calculated
in the approximation of the rotational isomer model $\mu_{calc, hind}$ with
three potential minima of the rotation around each C−C bond corresponding to two gauche and one trans configurations [225]. In
the latter case, the inductive effect was also taken into account by
the method of Smith et al. (Chapter V, Section 1).

As can be seen from the data of Table 29, the hypothesis of
free rotation does not permit an explanation of the observed dipole
moments. Conversely, the rotational isomerism model makes it
possible to harmonize the experimental and calculated values with
remarkable accuracy. This accuracy, however, is achieved by
selecting within the framework of the model used suitable values
of $\Delta E = E_{gauche} - E_{trans}$ of 0.4 for the intermediate and 0.34 kcal/mole
for the terminal links. Leonard et al. [226] have solved the problem of finding this magnitude theoretically on the basis of the experimental dipole moments given in Table 29 and the same rotational isomerism model, also taking into account, unlike Hayman
et al. [225], the correlation of rotations about different C−C bonds.
Leonard et al. [226] succeeded not only in harmonizing the calculated and experimental values of the dipole moments but also in
giving an extremely accurate prediction of their temperature dependence, which can be interpreted as the consequence of the adequacy for calculations of the model of the molecular structure.

The similar problem of verifying the correctness of the
model of the structure of regularly constructed molecules with a
long chain of atoms is exceptionally important in the study of the
structure of macromolecules in solutions. The results of determinations and the analysis of the dipole moments of a large series
of oligomers and polymers have been considered in detail in
Birshtein and Ptitsyn's monograph [227]. As one of the main conclusions arising from this analysis we may state that only the rotational isomerism model is capable of explaining the dipole moments of typical macromolecules observed experimentally.

Literature Cited

1. G. Briegleb and J. Kambeitz, Z. Phys. Chem., B27:11 (1934).
2. E. N. Di Carlo and C. P. Smyth, J. Am. Chem. Soc., 84:1128 (1962).
3. O. A. Osipov and V. I. Minkin, Handbook on Dipole Moments, Izd. Vysshaya shkola (1965).
4. E. Fischer and S. Schreiner, Ber., 92:938 (1959).
5. J. Trotter, Acta Cryst., 11:355 (1958).
6. O. A. Osipov, V. I. Minkin, and Zh. A. Tumakova, Zh. Strukt. Khim., 5:918 (1964).
7. V. A. Kogan, O. A. Osipov, V. I. Minkin, and M. I. Gorelov, Dokl. Akad. Nauk SSSR, 153:594 (1963).
8. O. A. Osipov, A. D. Garnovskii, and V. I. Minkin, Zh. Strukt. Khim., 8:919 (1967).
9. V. I. Minkin and L. E. Nivorozhkin, Zh. Organ. Khim., 1:1632 (1965).
10. A. N. Kiprianov and F. A. Mikhailenko, Abstracts of Papers at the Second All-Union Conference on the Chemistry of Five-Membered Nitrogen Heterocycles, Rostov-on-Don (1966), p. 162.
11. K. Higasi and S. Uyeo, J. Chem. Soc. Japan, 62:396 (1941).
12. J. G. M. Campbell, C. G. LeFevre, R. J. W. LeFevre, and E. S. Turner, J. Chem. Soc., 1938:404.
13. I. Yu. Kokoreva, Ya. K. Syrkin, et al., Dokl. Akad. Nauk SSSR, 166:155 (1966).
14. S. Hosoya, Acta Cryst., 16:310 (1963).
15. F. Pompa and A. Ripamondi, Ricerca Scientifica, Rend., 29:1316 (1959).
16. G. C. Hampson, Trans. Faraday Soc., 30:877 (1934).
17. I. J. Coop and L. E. Sutton, J. Chem. Soc., 1938:1269.
18. W. C. Hornig, F. Lautenschlayer, and G. F. Wright, Can. J. Chem., 41:1441 (1963).
19. E. D. Bergmann, E. Fischer, and B. Pullman, J. Chim. Phys., 48:356 (1951).
20. G. Ferguson and J. M. Robertson, Advances in Physical Organic Chemistry, Vol. 1, Academic Press, New York (1963), p. 218.
21. E. Bergman and A. Weizmann, Chem. Rev., 29:553 (1941).
22. A. W. Laubengayer and W. R. Rysz, Inorg. Chem., 4:1513 (1965).
23. N. J. Leonard and L. E. Sutton, J. Am. Chem. Soc., 70:1564 (1948).
24. I. R. Maxwell, S. B. Hendricks, and V. M. Mosley, J. Chem. Phys., 3:699 (1935).
25. J. Toussaint, Bull. Soc. Chim. Belge, 54:319 (1945).
26. A. I. Kitaigorodskii, Organic Crystal Chemistry, Izd. Akad. Nauk SSSR (1955).
27. V. W. Laurie, J. Chem. Phys., 34:291 (1961).
28. V. N. Vasil'eva, V. V. Perekalin, and V. G. Vasil'ev, Zh. Obshch. Khim., 31:2171 (1961).
29. K. B. Everard, L. Kumar, and L. E. Sutton, J. Chem. Soc., 1951:2807.
30. A. Demiel, J. Org. Chem., 27:3500 (1962).
31. J. A. Hugill, J. E. Coop, and L. E. Sutton, Trans. Faraday Soc., 34:1518 (1938).
32. J. D. Roberts, R. Armstrong, R. F. Trimble, and M. Burg, J. Am. Chem. Soc., 71:843 (1949).
33. J. E. Baldwin, J. Org. Chem., 29:1882 (1964).
34. G. S. Hartley and R. J. W. LeFevre, J. Chem. Soc., 1939:531.
35. M. Frankel, R. Wolovsky, and E. Fischer, J. Chem. Soc., 1955:3441.
36. W. West and R. B. Killingsworth, J. Chem. Phys., 6:1 (1938).

37. C. G. Overberger, J. Am. Chem. Soc., 85:2752 (1963).

38. R. J. W. LeFevre and J. Northcott, J. Chem. Soc., 1953:867.

39. R. J. W. LeFevre and J. Northcott, J. Chem. Soc., 1949:944.

40. V. I. Minkin, O. A. Osipov, and V. A. Kogan, Dokl. Akad. Nauk SSSR, 145:336 (1962).

41. V. de Gaouck and R. J. W. LeFevre, J. Chem. Soc., 1938:741; 1939:1392.

42. F. Feichtmayr and F. Würstlin, Chem. Abs., 55:24172 (1961).

43. T. Kubota, M. Yamakawa, and Y. Mori, Bull. Chem. Soc. Japan, 36:1552 (1963).

44. L. E. Sutton and T. W. J. Taylor, J. Chem. Soc., 1933:2190.

45. W. Theilacker and K. Fauser, Ann. Chemie, 539:103 (1939).

46. O. Exner, Collection Czech. Commun., 30:652 (1965).

47. F. Feichtmayr and F. Würstlin, Ber. Bunsenges. Physik, Chem., 67:434 (1963).

48. J. F. King and T. Durst, J. Am. Chem. Soc., 85:2676 (1963).

49. H. J. Becker and H. Diehl, Ber., 98:526 (1965).

50. R. S. Mulliken, Rev. Mod. Phys., 14:265 (1942).

51. R. A. Beaudet, J. Am. Chem. Soc., 88:1390 (1966).

52. A. E. Lutskii, V. T. Alekseeva, and B. P. Kondratenko, Zh. Fiz. Khim., 35:839 (1961).

53. A. A. Antony and C. P. Smyth, J. Am. Chem. Soc., 86:156 (1964).

54. W. Kauzmann, Quantum Chemistry, Academic Press, New York (1957).

55. J. M. Lehn and G. Ourisson, Bull Soc. Chim. France, 1963:1113.

56. L. M. Nazarova and Ya. K. Syrkin, Izv. Akad. Nauk SSSR, Ser. Khim., 1949:35.

57. K. N. Kovalenko, V. I. Minkin, Z. N. Nazarova, and D. V. Kazachenko, Zh. Obshch. Khim., 32:549 (1962).

58. H. Lumbroso and P. Pastour, Compt. Rend., 261:1279 (1965).

59. A. Marinangeli, Ann. Chim. (Rome), 44:219 (1954).

60. J. Barassin, Ann. Chim. (Paris), 8:666 (1963).

61. V. I. Minkin, Yu. A. Zhdanov, A. D. Garnovskii, and I. D. Sadekov, Zh. Fiz. Khim., 40:657 (1966).

62. R. Crigg and M. V. Sargent, Tetrahedron Letters, 1965:1381.

63. G. J. Karabatsos and F. M. Vane, J. Am. Chem. Soc., 85:3886 (1963).

64. K. Hafner, H. Kramer, H. Musso, G. Ploss, and G. Schulz, Ber., 97:2066 (1964).

65. P. H. Cureton, C. G. LeFevre, and R. J. W. LeFevre, J. Chem. Soc., 1961:4447.

66. R. J. W. LeFevre and P. J. Stiles, J. Chem. Soc., B:420 (1966).

67. S. V. Tsukerman, A. I. Artemenko, and V. F. Lavrushin, Zh. Obshch. Khim., 34:3591 (1964).

68. S. V. Tsukerman, Kuo Sheng Chang, and V. F. Lavrushin, Zh. Fiz. Khim., 40:160 (1966).

69. V. Baliah and M. Uma, Tetrahedron, 19:455 (1963).

70. W. F. Anzilotti and B. C. Curran, J. Am. Chem. Soc., 65:607 (1943).

71. J. W. Williams, Phys. Z., 29:683 (1928).

72. J. S. Dryden and B. J. Meakins, Proc. Phys. Soc., 69B:252 (1956).

73. H. Lumbroso and G. Palamidessi, Bull. Soc. Chim. France, 1965:3150.

74. H. Lumbroso and R. Passerini, Bull. Soc. Chim. France, 1957:311.

75. L. E. Sutton, Determination of Organic Structures by Physical Methods, New York (1955), p. 272.

76. R. J. W. LeFevre, A. Sundaram, and K. M. S. Sundaram, J. Chem. Soc., 1963:4447.
77. V. Balasubramaniyan, Chem. Rev., 66:567 (1966).
78. G. F. Longster and E. E. Walker, Trans. Faraday Soc., 49:228 (1954).
79. R. Huisgen, Angew. Chem., 69:341 (1957).
80. R. F. Curl, J. Chem. Phys., 30:1529 (1959).
81. T. Miyazawa, Bull. Chem. Soc. Japan, 34:691 (1961).
82. C. P. Smyth, Dielectric Behavior and Structure, McGraw Hill, New York (1955).
83. Y. Hirabayashi, Bull. Chem. Soc. Japan, 38:175 (1965).
84. H. Lumbroso and P. Reynaud, Compt. Rend., 262:1739 (1966).
85. C. M. Lee and W. D. Kumler, J. Am. Chem. Soc., 83:4586 (1961); 84:565, 571 (1962).
86. A. C. Littlejohn and J. W. Smith, J. Chem. Soc., 1953:2456; 1954:2552.
87. O. Bastiansen, Acta Chem. Scand., 3:408 (1949); 4:926 (1950).
88. Tables of Interatomic Distances and Configuration in Molecules and Ions, Publ. No. 11, Chem. Soc., London (1958).
89. A. Allmenningen and O. Bastiansen, Kgl., Norske Vid. Selsk. Skr., No. 4 (1958).
90. G. C. Hampson and A. Weissberger, J. Am. Chem. Soc., 58:2111 (1936).
91. T. Sato, Nippon Kagaku Zasshi, 33:501 (1950).
92. B. J. Kurland and W. B. White, J. Am. Chem. Soc., 86:1877 (1964).
93. D. L. Smare, Acta Cryst., 1:150 (1948).
94. F. Fowwether and A. Hargreaves, Acat Cryst., 3:81 (1950).
95. H. Weller-Feilchenfeld, E. D. Bergmann, and A. Hirschfeld, Tetrahedron Letters, 1965:4129.
96. E. D. Bergmann, H. Weller-Feilchenfeld, A. Heller, C. Britzmann, and A. Hirschfeld, Tetrahedron, Suppl. 7:349 (1966).
97. K. Higasi and S. Uyeo, Bull. Chem. Soc. Japan, 14:87 (1939).
98. H. Lumbroso and G. Montando, Bull. Soc. Chim. France, 1964:2119.
99. C. W. N. Cumper, J. F. Read, and A. I. Vogel, J. Chem. Soc., 1965:5860.
100. R. S. Tsekhanskii and L. I. Vinogradov, Zh. Obshch. Khim., 32:3802 (1962).
101. R. J. W. LeFevre, A. Sundaram, and K. M. S. Sundaram, Bull. Chem. Soc. Japan, 35:690 (1962).
102. M. J. Aroney and R. J. W. LeFevre, J. Chem. Soc., 1963:1167, 4450; 1965:5810.
103. R. J. W. LeFevre and J. D. Saxby, J. Chem. Soc., 1966 B:1064.
104. V. A. Izmail'skii and E. A. Smirnov, Zh. Obshch. Khim., 26:3042 (1956).
105. V. I. Minkin, Yu. A. Zhdanov, E. A. Medyantzeva, and Yu. A. Ostroumov, Tetrahedron, 23:3651 (1967).
106. D. J. W. Bullock, C. W. N. Cumper, and A. I. Vogel, J. Chem. Soc., 1965:5316.
107. C. F. Wilcox, J. Am. Chem. Soc., 82:414 (1960).
108. F. V. Brutcher and W. Bauer, J. Am. Chem. Soc., 84:2233 (1962).
109. C. Altona, H. R. Buys, and E. Havinga, Rec. Trav. Chim., 85:973, 983 (1966).
110. E. J. Corey and R. A. Sneen, J. Am. Chem. Soc., 77:2505 (1955).
111. K. Altenburg, Z. Chem., 6:477 (1966).
112. R. J. W. LeFevre, J. Chem. Soc., 1965:3701.
113. A. K. Bose, M. S. Manhas, and E. P. Malinowski, J. Am. Chem. Soc., 85:2795 (1963).
114. N. S. Zefirov, G. P. Krutetskaya, L. G. Prikazchikova, and Yu. K. Yur'ev, Zh. Obshch. Khim., 35:1687 (1965).

115. D. D. Tanner and T. S. Gilman, J. Am. Chem. Soc., 85:2892 (1963).

116. R. Riemschneider and F. D. Grabitz, Bochu Kagaku, 26:99 (1961).

117. R. Riemschneider and V. Wuscherpfennig, Z. Naturforsch., 17b:585 (1962).

118. G. F. Wright, Zh. Vsesoyuzn. Khim. Obshchestva, 7:300 (1962).

119. F. Lautenschlaeger and G. F. Wright, Can. J. Chem., 41:863 (1963).

120. J. B. Lambert and J. D. Roberts, J. Am. Chem. Soc., 85:3710 (1963); 87:3884 (1965).

121. K. B. Wiberg and G. M. Lampman, J. Am. Chem. Soc., 88:4429 (1966).

122. W. G. Rothschild and B. P. Dailey, J. Chem. Phys., 36:2931 (1962).

123. B. A. Arbuzov, O. N. Nuretdinova, and A. N. Vereshchagin, Dokl. Akad. Nauk SSSR, 172:591 (1967).

124. E. Eliel, Stereochemistry of Carbon Compounds, McGraw-Hill, New York (1962).

125. E. L. Eliel, N. L. Allinger, S. J. Angyal, and G. A. Morrison, Conformational Analysis, Interscience, New York (1965).

126. F. V. Brutcher, T. Roberts, S. J. Barr, and N. Pearson, J. Am. Chem. Soc., 81:4915 (1959).

127. B. A. Arbuzov and L. K. Yuldasheva, Izv. Akad. Nauk SSSR, Ser. Khim., 1734 (1962).

128. S. Winstein and M. J. Holness, J. Am. Chem. Soc., 77:5562 (1955).

129. A. N. Vereshchagin and S. G. Vul'fson, Teoret. Éksperim. Khim., Akad. Nauk UkrSSR, 1:306 (1965).

130. H. R. Nace and R. H. Nealey, J. Am. Chem. Soc., 88:65 (1966).

131. H. J. Geise, A. Tieleman, and E. Havinga, Tetrahedron, 22:183 (1966).

132. H. J. Hageman and E. Havinga, Tetrahedron, 22:2271 (1966).

133. W. Hückel and H. Waiblinger, Ann. Chemie, 666:17 (1963).

134. P. Klaboe, J. J. Lothe, and K. Lunde, Acta Chem. Scand., 11:1677 (1957).

135. I. Miyagawa, Y. Morino, and R. Riemschneider, Bull. Chem. Soc. Japan, 27:177 (1954).

136. R. Riemschneider and F. Scheppler, Monatsh., 86:548 (1955).

137. C. Quivoron and J. Neel, J. Chim. Phys., 61:554 (1964).

138. Y. Morino and I. Miyagawa, Bochu Kagaku, 15:181 (1950).

139. T. Shimozawa, Bull. Chem. Soc. Japan, 28:384 (1955).

140. E. Gey, Z. Chem., 7:35 (1967).

141. J. Allinger and N. L. Allinger, Tetrahedron, 2:64 (1958).

142. N. L. Allinger, J. Allinger, L. A. Freiberg, R. F. Czaja, and N. A. Le Bel, J. Am. Chem. Soc., 82:5876 (1960).

143. N. L. Allinger and H. M. Blattner, J. Org. Chem., 27:1523 (1962).

144. W. D. Kumler and A. C. Huitric, J. Am. Chem. Soc., 78:3369 (1956).

145. P. Mauret and J. Petrissans, Compt. Rend., 254:3362 (1962).

146. N. L. Allinger, J. Allinger, and N. A. Le Bel, J. Am. Chem. Soc., 82:2926 (1960).

147. C. Y. Chen and R. J. W. LeFevre, J. Chem. Soc., 1965:3700.

148. N. L. Allinger, J. Allinger, M. A. DaRooge, and S. Greenberg, J. Org. Chem., 27:4603 (1962).

149. J. Lehn, J. Levisalles, and G. Ourisson, Tetrahedron Letters, 1961:682.

150. N. L. Allinger, J. Allinger, and M. A. DaRooge, J. Am. Chem. Soc., 86:4061 (1964).

151. N. L. Allinger and M. A. DaRooge, Tetrahedron Letters, 1961:676.

152. C. G. LeFevre and R. J. W. LeFevre, J. Chem. Soc., 1935:1696.

153. N. L. Allinger, H. M. Blattner, L. A. Freiberg, and F. M. Karkowski, J. Am. Chem. Soc., 88:2999 (1966).

154. M. T. Rogers and J. M. Caneon, J. Phys. Chem., 65:1417 (1961).

155. M. Aroney and R. J. W. LeFevre, J. Chem. Soc., 1958:3002.

156. N. L. Allinger, J. G. D. Carpenter, and F. M. Karkowski, J. Am. Chem. Soc., 87:1232 (1965).

157. R. J. Bishop, L. E. Sutton, D. Dineen, R. A. Y. Jones, and A. R. Katritzky, Proc. Chem. Soc., 1964:257.

158. M. J. Aroney, C. Y. Chen, R. J. W. LeFevre, and A. N. Singh, J. Chem. Soc., 1966 B:98.

159. J. M. Eckert and R. J. W. LeFevre, J. Chem. Soc., 1962:3991.

160. R. J. Bishop, G. Fodor, A. R. Katritzky, F. Soti, and L. E. Sutton, J. Chem. Soc., 1966 C:74.

161. M. J. Aroney, C. Y. Chen, R. J. W. LeFevre, and J. D. Saxby, J. Chem. Soc., 1964:4269.

162. B. A. Arbuzov, G. G. Butenko, and A. N. Vereshchagin, Dokl. Akad. Nauk SSSR, 172:1323 (1967).

163. K. E. Calderbank and R. J. W. LeFevre, J. Chem. Soc., 1949:199.

164. J. Crossley, A. Holt, and S. Walker, Tetrahedron, 21:3141 (1965).

165. C. W. N. Cumper and A. I. Vogel, J. Chem. Soc., 1959:3521.

166. B. A. Arbousow [Arbusov], Bull. Soc. Chim. France, 1960:1311.

167. T. Urbanski, Roczn. Chem., 25:183 (1951).

168. A. V. Bogatskii, Yu. Yu. Samitov, and N. A. Garkovik, Zh. Organ. Khim., 2:1335 (1966).

169. T. Urbanskii, Zh. Vsesoyuzn. Khim. Obshchestva, 7:396 (1962).

170. H. T. Kalff and E. Havinga, Rec. Trav. Chim., 81:282 (1962).

171. H. T. Kalff and E. Havringa, Rec. Trav. Chim., 85:467 (1966).

172. B. A. Arbuzov and T. G. Shavsha, Dokl. Akad. Nauk SSSR, 69:41 (1949).

173. D. G. Hellier, J. G. Tillett, U. F. Van Woerden, and R. F. M. White, Chem. Ind. (London), 1963:1956.

174. K. Hayasaki, J. Chem.Soc. Japan, 76:284 (1955).

175. R. A. Florentine and G. Miller, J. Am. Chem. Soc., 81:5103 (1959).

176. H. Bradford Thomson, L. P. Kuhn, and M. Inatome, J. Phys. Chem., 68:421 (1964).

177. R. Borsdorf, R. Heckel, and M. Mühlstadt, Z. Chem., 5:25 (1965).

178. N. L. Allinger and S. E. Hu, J. Am. Chem. Soc., 83:1664 (1961).

179. N. J. Leonard, T. W. Milligan, and T. L. Brown, J. Am. Chem. Soc., 82:4075 (1960).

180. M. Baron, O. B. Mandriola, and J. F. Westerkamp, Can. J. Chem., 41:1893 (1963).

181. W. D. Kumler, A. Lewis, and J. Meinwald, J. Am. Chem. Soc., 83:4591 (1961).

182. C. F. Wilcox, J. G. Zajacek, and M. F. Wilcox, J. Org. Chem., 30:2621 (1965).

183. B. Waegell and C. F. Jefford, Bull. Soc. Chim. France, 1964:844.

184. W. D. Kumler, R. Boikess, R. Bruck, and S. Winstein, J. Am. Chem. Soc., 86:3126 (1964).

185. M. D. Harmony and K. Cox, J. Am. Chem. Soc., 88:5049 (1966).

186. B. A. Arbuzov, A. N. Vereshchagin, and S. G. Vul'fson, Izv. Akad. Nauk SSSR, Ser. Khim., 1965:155.

187. B. A. Arbuzov and A. N. Vereshchagin, Izv. Akad. Nauk SSSR, Ser. Khim., 1731 (1966).

188. B. A. Arbuzov, A. N. Vereshchagin, and É. S. Batyeva, Izv. Akad. Nauk SSSR, Ser. Khim., 2083 (1966).

189. F. Caplan and H. Conroy, J. Org. Chem., 28:1593 (1963).

190. B. A. Arbuzov, A. N. Vereshchagin, and É. S. Batyeva, Izv. Akad. Nauk SSSR, Ser. Khim., 10 (1967).

191. T. L. Brown, D. Y. Curtin, and R. R. Fraser, J. Am. Chem. Soc., 80:4339 (1958).

192. L. Hörner and W. Durckheimer, Ber., 95:1219 (1962).

193. N. A. Le Bel, J. Am. Chem. Soc., 82:623 (1960).

194. A. N. Vereshchagin, Author's Abstract of Candidate's Thesis, Kazan' State University, Kazan' (1964).

195. A. N. Vereshchagin and S. G. Vul'fson, Teoret. Éksperim. Khim., 1:305 (1965).

196. B. A. Arbuzov and A. N. Vereshchagin, Izv. Akad. Nauk SSSR, Ser. Khim., 1004 (1964).

197. A. N. Vereshchagin and B. A. Arbuzov, Izv. Akad. Nauk SSSR, Ser. Khim., 35 (1965).

198. M. Piconays, Bull. Soc. Chim. France, 1964:465.

199. N. L. Allinger and W. Szkryabalo, J. Org. Chem., 27:722 (1962).

200. T. Fujita, J. Am. Chem. Soc., 79:2471 (1957).

201. B. Krishna, B. B. L. Saxena, and S. C. Srivastava, Tetrahedron, 22:1597 (1966).

202. A. N. Shidlovskaya, Ya. K. Syrkin, and I. N. Nazarov, Izv. Akad. Nauk SSSR, Ser. Khim., 241 (1958).

203. V. I. Minkin and M. I. Gorelov, Zh. Obshch. Khim., 33:647 (1963).

204. N. L. Allinger and M. A. DaRooge, Tetrahedron Letters, 1961:682.

205. M. Aroney, L. N. L. Chia, and R. J. W. LeFevre, J. Chem. Soc., 1961:4144.

206. A. Weissberger and R. Sängewald, Z. Phys. Chem., B9:133 (1930).

207. F. Eisenlohr and L. Hill, Z. Phys. Chem., B36:30 (1937).

208. G. Drefahl, G. Heublein, and D. Voigt, J. Prakt. Chem., 23:157 (1964).

209. S. Mizushima, The Structure of Molecules and Internal Rotation [Russian translation], IL, Moscow (1957), Chapters 1 and 2.

210. J. Ainsworth and J. Karle, J. Chem. Phys., 20:245 (1952).

211. R. J. W. LeFevre and B. J. Orr, Australian J. Chem., 17:1048 (1964).

212. R. J. W. LeFevre, C. D. L. Ritchie, and P. J. Stiles, Chem. Comm., 1966:846.

213. M. J. Aroney, R. J. W. LeFevre, and J. D. Saxby, J. Chem. Soc., 1966B:414.

214. P. A. Akishin, L. V. Vilkov, and N. P. Sokolova, Izv. Sibirsk. Otd. Akad. Nauk SSSR, 5:59 (1960).

215. M. C. Planje, L. H. Toneman, and G. Dalinga, Rec. Trav. Chim., 84:232 (1965).

216. R. G. Jones and W. J. Orville-Thomas, J. Chem. Soc., 1964:692.

217. H. Lumbroso and D. Lauransan, Bull. Soc. Chim. France, 1959:513.

218. C. T. Mortimer, H. Spedding, and H. Springall, J. Chem. Soc., 1957:188.

219. C. Beguin and T. Gaumann, Helv. Chim. Acta, 41:1971 (1957).

220. B. A. Arbuzov and T. G. Shavsha, Dokl. Akad. Nauk SSSR, 68:515, 859 (1949).

221. C. W. N. Cumper, A. Melnikoff, and A. I. Vogel, J. Chem. Soc., 1966A:323.

222. K. Matsumura, Bull. Chem. Soc. Japan, 35:801 (1962).

223. K. Yamasaki, A. Kotera, M. Yokoi, and Y. Ueda, J. Chem. Phys., 18:1414 (1950).
224. I. I. Zemlyanskii, I. I. Gol'dshtein, and E. N. Gur'yanova, Dokl. Akad. Nauk SSSR, 156:131 (1964).
225. H. J. G. Hayman and I. Eliezer, J. Chem. Phys., 28:890 (1958); 35:644 (1961).
226. W. J. Leonard, R. L. Jernigan, and P. J. Flory, J. Chem. Phys., 43:2256 (1965).
227. T. M. Birshtein and O. B. Ptitsyn, The Comformation of Macromolecules, Izd. Nauka, Moscow (1964), Chapters 1, 5, and 6.

Chapter V

Dipole Moments and the Electronic Structure of Organic Compounds

The dipole moment of a molecule is a direct characteristic of its electronic configuration. Perturbation of the latter under the influence of certain structural factors such as the introduction of substituting groups into the molecule is directly reflected in the magnitude of the dipole moment. Various structural rearrangements of the molecule are connected with a redistribution of the electron density by two mechanisms: the induction mechanism and the conjugation mechanism. The first of them is universal while the second acts only in the case of unsaturated molecules. In the present chapter we shall consider methods for the quantitative evaluation of these and of a number of specific effects on the distribution of the electron density in molecules of various types based on dipole moment data.

1. Induction Effect

Two main approaches to the quantitative evaluation of the influence of the induction effect of polar groups on the magnitude of dipole moments exist. The first of them is based on a consideration of the molecule as a system of several fragments polarized in the field created by the dipole of the polar group. The second proposes to take account of the effects of the polarization of each bond present in the molecule.

In each of these methods, the calculation takes into consideration only one of the components of the induction effect. In the first case, this is the effect of the field, which exerts a considerable influence on the total magnitude of the induced moments. In this, the effect of the successive polarization of the σ-bonds, i.e., the σ-induction effect, is ignored. The calculations comprising an evaluation of the polarization of each electron pair of the molecule take into account the σ-induction effect almost alone.

In spite of such large differences in the calculation of induced moments, the two approaches lead to results agreeing fairly well with one another. In this connection it is appropriate to draw a parallel with evaluations of the inductive influences on the reactivity of organic compounds [1]. Although a considerable number of data indicate, in the main, the field nature of the induction effect, the σ* constants of polar substituents in the aliphatic series do not depend on the conformation of the substituent and are determined only by the sequence of bonds connecting them to the reaction center.

Calculation of the Induced Dipole Moments Considering

the Molecule as a System of Polarized Fragments

It follows from classical electrostatics that a permanent dipole placed at the origin of coordinates of a system creates at some point at distance r an electric field E the components of which along axes of coordinates in the plane containing this point and the dipole are given by the equations [2]

$$
\begin{aligned}
E_x &= \frac{\mu_x (3 \cos^2 \theta - 1)}{\varepsilon_a r^3} + \frac{3\mu_y \sin \theta \cos \theta}{\varepsilon_a r^3} \\
E_y &= \frac{3\mu_x \sin \theta \cos \theta}{\varepsilon_a r^3} + \frac{\mu_y (3 \sin^2 \theta - 1)}{\varepsilon_a r^3}
\end{aligned}
\tag{V.1}
$$

where μ_x and μ_y are the components of the vector $\vec{\mu}$; ε_a is the dielectric constant of the medium, and θ is the angle between the axis of the dipole and the radius vector of the point under consideration.

The use of the equations given for calculating induced dipole moments requires a number of assumptions: a) concerning the position of the point of localization of the polarizing dipole μ, since relation (V.1) was derived for the case of a point dipole, b) concerning the dielectric constant of the medium between the point of localization of the dipole and the point at which the additional moment is induced. If these assumptions are introduced, the components of the induced dipole μ_i are calculated by means of the equations

$$\mu_{ix} = \alpha E_x; \qquad \mu_{iy} = \alpha E_y \qquad (V.2)$$

where α is the polarizability of the molecule at the point considered.

The equation of the field of a dipole was somewhat refined by Frank [3]. According to Frank, the components of the induced dipole are expressed as

$$\mu_{ix} = \frac{\mu \alpha \, (\varepsilon_b + 2)}{3 \varepsilon_a r^3} \, (3 \cos^2 \theta - 1)$$

$$\mu_{iy} = \frac{\mu \alpha \, (\varepsilon_b + 2)}{3 \varepsilon_a r^3} \cdot 3 \sin \theta \cos \theta \qquad (V.3)$$

where ε_b is the dielectric constant of the polarized group.

Equation (V.3) is most frequently used to evaluate the induced moment. It is generally assumed that the point of localization of the dipole of the polar group is at the "point of contact" of the atoms forming the dipole, i.e., at the distance of their covalent radii along the line joining them, and ε_b is taken as equal to the dielectric constant of benzene, naphthalene, or a saturated hydrocarbon, depending on the class to which the compound under study belongs. It is difficult to make a reliable assumption concerning the magnitude of ε_a. Smallwood and Herzfeld [2] take ε_a as equal to the dielectric constant of a vacuum. The assumption of the equality of ε_a and ε_b made by Smith [4] appears to be more reliable.

The main advantage of the method of calculating induced moments that has been given is the possibility of taking into account the mutual orientation of the interacting dipoles in space. The role of this effect is shown by the fact that the dipole moments of bonds in molecules with several strong dipoles depend fundamentally on the relative disposition of these dipoles [5, 6]. This effect, of course, cannot be described by methods considering only the successive polarization of σ-bonds.

However, the use of equations (V.1) and (V.3) to calculate induced dipoles is possible only for comparatively large distances between them (r > 2 Å), since for smaller values of r they do not reflect the distribution of the potential of the field of a point dipole [2]. In view of this, calculations by the method given are carried out mainly for derivatives of aromatic compounds where the condition r > 2 Å is satisfied even for ortho-disubstituted derivatives. In the aliphatic series, for example in the calculation of the moments of halogen-substituted methanes, the use of equations (V.1) and (V.3) does not lead to results agreeing with the experimental figures.

The considerable degree of arbitrariness that must be accepted in the selection of the initial data for calculating induced dipoles has led to a whole series of attempts to refine the method of calculation. In particular, Groves and Sugden [7] have used graphical integration to eliminate the inaccuracies due to the selection of the point of localization of the dipole. This work did not, however, lead to clarity in the determination of ε_a, and calculation by this method is too tedious.

Detailed reviews of various modifications of the methods of calculating induced dipole moments based on developments of the methods discussed above have been given by Smith [4] and Vereshchagin [8].

Calculation of Induced Dipole Moments taking the Effects of the Polarization of Each Individual Bond into Account

The alternative procedure for the theoretical treatment of the influence of the induction effect on dipole moments proposes a consideration of the polarization of each valence electron pair of the molecule. Such an approach was first suggested by Smyth [9] and Remick [10], but a more general method suitable for concrete calculations has been developed by Smith et al. [11, 12].

According to (V.2) the dipole induced in a bond $a-b$ through the polarization of the electron cloud of the two bonding electrons can be represented in the form

$$- Q_a^b R_{a-b} = \left(b_l\right)_{a-v} \left(\frac{Z_a e}{R_a^2} - \frac{Z_b e}{R_b^2}\right) \qquad (V.4)$$

where Q_a^b is the charge arising on atom a as a result of the polarization of the $a-b$ bond, R_{a-b} is the length of the $a-b$ bond, $(b_l)_{a-b}$ is the polarizability along the line of the bond, Z_a and Z_b are the effective nuclear charges of the atoms a and b, R_a and R_b are their covalent radii, and e is the electronic charge.

The effective nuclear charges are defined as

$$Z_a = Z_a^0 + \frac{S_a}{e}\,\varepsilon_a$$
$$Z_b = Z_b^0 + \frac{S_b}{e}\,\varepsilon_b \tag{V.5}$$

where Z_a^0 is the nuclear charge of the neutral atom a, S_a is the Slater screening constant [13], and ε_a is the total charge on atom a, which is equal to the sum of all the Q_a^b. The symbols relating to atom b have analogous meanings.

Substituting (V.5) in (V.4), we obtain

$$Q_a^b = -\frac{(b_l)_{a-b}e}{R_{a-b}}\left[\frac{Z_a^0}{R_a^2} - \frac{Z_b^0}{R_b^2} + \frac{S_a\varepsilon_a}{R_a^2 e} - \frac{S_b\varepsilon_b}{R_b^2 e}\right] \tag{V.6}$$

Let us introduce the following symbols:

$$\alpha_{a-b} = -\frac{(b_l)_{a-b}e}{R_{a-b}}\left(\frac{Z_a^0}{R_a^2} - \frac{Z_b^0}{R_b^2}\right); \quad \beta_b^a = \frac{S_b(b_l)_{a-b}}{R_{a-b}R_b^2}; \quad \beta_a^b = \frac{S_a(b_l)_{a-b}}{R_{a-b}R_a^2} \tag{V.7}$$

Then (V.6) can be written in the form

$$Q_a^b = \alpha_{a-b} + \beta_b^a\varepsilon_b - \beta_a^b\varepsilon_a \tag{V.8}$$

The values of the parameters β_b^a and β_a^b can easily be calculated if those of $(b_l)_{a-b}$ are known (Table 30). In the literature [11, 12], for $(b_l)_{a-b}$ values borrowed mainly from earlier sources [14, 15] are taken. At the present time, more detailed and accurate data are available [16]. However, in view of the empirical nature of the calculations and the approximations upon which the initial formula (V.4) is based,* the magnitudes $(b_l)_{a-b}$ can be regarded as parameters, from which it follows that the replacement

---

* Formula (V.4) is correct if the following conditions are satisfied: a) the moment of the bond is defined in terms of the magnitude of the charges localized on the atomic nuclei, and b) atoms a and b have a spherical electric potential. We may also note that possible inaccuracies in the value of b_l in formulas (V.7) are compensated by the empirical selection of γ_{a-b} according to (V.9).

TABLE 30. The Parameters β_b^a, β_a^b, and β_{a-b}^- according to the Literature [4, 11, 12]

Bond a-b	$b_l \cdot 10^{24}$, cm^3	R_a, Å	R_b, Å	β_a^b	β_b^a	β_{a-b}^-
C—C	1.12	0.771	0.771	0.43	0.43	0
C—F	0.96	0.771	0.64	0.401	0.581	0.25
C—Cl	3.67	0.771	0.99	1.23	0.744	0.71
C—Br	5.04	0.771	1.14	1.55	0.710	0.91
C—I	8.09	0.771	1.33	1.27	0.762	1.29
C—H	0.79	0.771	0.30	0.434	2.46	0.13
C—O	0.84	0.771	0.66	0.346	0.472	0.23
C—N	0.86	0.771	0.70	0.344	0.418	0.24
N—H	0.58	0.70	0.30	0.414	1.94	0.14
S—H	2.30	1.04	0.30	0.555	5.72	0.08
C=O	1.99	0.665	0.55	1.30	1.90	—
C=C	2.86	0.665	0.665	1.70	1.70	—
C=S	7.57	0.665	0.94	3.73	1.87	—
N≡N	2.43	0.547	0.547	2.60	2.60	—
C≡N	3.1	0.602	0.547	2.61	3.16	—
C≡C	3.54	0.602	0.602	2.84	2.84	—
C_{Ar}—C_{Ar}	2.25	0.695	0.695	1.17	1.17	—

of the old values of the polarizability by the new ones does not substantially affect the accuracy of the calculations. The values of α_{a-b} can also be calculated by means of (V.7), but they prove to be very sensitive to the choice of b_l and R_a, R_b. Consequently, α_{a-b} is best evaluated empirically from the magnitude of the dipole moments of some standard compounds.

Let us introduce the symbols

$$Y_{a-b} = \frac{a_{a-b}}{1+\beta_a^b}; \quad \beta_{a-b} = \frac{\beta_b^a}{1+\beta_b^a} \tag{V.9}$$

If a is an atom of H, F, Cl, Br, I, O (in the CO group), etc., chemically bound to only one other atom b, while b is a carbon atom, then $Q_a^b = \varepsilon_a$ and according to (V.8) and (V.9)

$$\varepsilon_X = Y_{X-C} + \beta_{X-C}\varepsilon_C \tag{V.10}$$

Using the values of β_a^b from Table 30, we obtain

$$\beta_{H\perp C} = 0.13; \quad \beta_{F-C} = 0.25; \quad \beta_{Cl-C} = 0.71; \quad \beta_{Br-C} = 0.91; \quad \beta_{I-C} = 1.29 \tag{V.11}$$

For a molecule of the type CH_3X where X is halogen, we have

$$\varepsilon_X = Y_{X-C} + \beta_{X-C}\varepsilon_C$$
$$\varepsilon_H = Y_{H-C} + \beta_{H-C}\varepsilon_C \tag{V.12}$$
$$\varepsilon_X + \varepsilon_C + 3\varepsilon_H = 0$$

TABLE 31. Values of γ_{X-C} according to
Smith et al. [11, 12]

Bond	$\gamma_{X-C} \cdot 10^{10}$ (CGSE) for various values of μ_{C-H}		
	$\mu_{C^--H^+}=0.3\,D$	$\mu_{C-H}=0$	$\mu_{C^+H^-}=0.3\,D$
H—C	0.418	0	—0.418
F—C	—0.944	—1.44	—1.93
Cl—C	—0.584	—1.49	—2.40
Br—C	—0.354	—1.44	—2.53
I—C	0.081	—1.42	—2.82

The combined solution of equations (V.12) gives the values of the total charges on each atom in the molecule of CH_3X in the form of functions of the parameters γ and β:

$$\varepsilon_H = \frac{(1+\beta_{X-C})\gamma_{H-C}-\beta_{H-C}\gamma_{X-C}}{1+\beta_{X-C}+3\beta_{H-C}}$$

$$\varepsilon_C = \frac{-\gamma_{X-C}-3\gamma_{H-C}}{1+\beta_{X-C}+3\beta_{H-C}} \qquad (V.13)$$

$$\varepsilon_X = \frac{(1+3\beta_{H-C})\gamma_{X-C}-3\beta_{X-C}\gamma_{H-C}}{1+\beta_{X-C}+3\beta_{H-C}}$$

If it is assumed, in agreement with the experimental data [17], that all the angles in the halomethanes correspond approximately to the tetrahedral angle, then

$$\mu_{CH_3X} = -\varepsilon_X R_{C-X} + \varepsilon_H R_{C-H}$$

The substitution of expression (V.13) in (V.14) gives

$$\mu_{CH_3X}(1+\beta_{X-C}+3\beta_{H-C}) = \gamma_{H-C}[3\beta_{X-C}R_{C-X}+(1+\beta_{X-C})R_{C-H}]$$
$$-\gamma_{X-C}[(1+3\beta_{H-C})R_{C-X}+\beta_{H-C}R_{C-H}] \qquad (V.15)$$

Equation (V.15) is used to calculate γ_{X-C}. The value of γ_{H-C} is determined from the moment of the H—C bond in the following way. For methane, according to (V.13) where X = H:

$$\gamma_{H-C} = (1+4\beta_{H-C})\varepsilon_H = 1.52\varepsilon_H \qquad (V.16)$$

Taking into account, moreover, the fact that ε_H can be determined from the moment of the C—H bond and the R_{C-H} distance of 1.09 Å

$$\mu_{C-H} = -\varepsilon_H R_{C-H}$$

it is easy to calculate γ_{H-C}. These values therefore depend both on the magnitude and on the direction of the moment of the $C-H$ bond. Since this magnitude is fairly indeterminate, Smith et al. [11, 12] calculated the values of γ_{H-C} and, from (V.15), γ_{X-C} for several of the most realistic values of μ_{C-H}. These values are given in Table 31.

In spite of the substantial variations in the value of γ_{X-C} according to the value of μ_{C-H} taken, the calculated values of the dipole moments of different haloalkanes depend little on what particular value of γ_{X-C} is used. In a later paper, Smith and Mortensen [12] recommended the use in calculations of the value of γ_{X-C} obtained with $\mu_{C-H} = 0$.

As an example of the calculation of dipole moments by the method of Smith et al., let us calculate the dipole moment of dichloromethane. In this case, equations (V.12) are replaced by the relations

$$\varepsilon_{Cl} = \gamma_{Cl-C} + \beta_{Cl-C}\varepsilon_C$$
$$\varepsilon_H = \gamma_{H-C} + \beta_{H-C}\varepsilon_C$$
$$\varepsilon_C + 2\varepsilon_{Cl} + 2\varepsilon_H = 0$$

On substituting the values of γ from Table 31 and those of β from Table 30, we obtain the following magnitudes for the full charges on the atoms (in CGSE units $\cdot 10^{-10}$):

$$\varepsilon_C = 1.11; \quad \varepsilon_H = 0.145; \quad \varepsilon_{Cl} = -0.70$$

Having located the origin of coordinates at the position of the central carbon atom and operating as described in Chapter III, we obtain

$$\sum \mu_x = 0; \quad \sum \mu_z = 0$$
$$\sum \mu_y = -2\varepsilon_{Cl}R_{C-Cl}\cos\frac{\varphi_1}{2} - 2\varepsilon_H R_{C-H}\cos\frac{\varphi_2}{2}$$

where R_{C-Cl} and R_{C-H} are the lengths of the $C-Cl$ and $C-H$ bonds, equal, respectively, to 1.78 and 1.09 Å, φ_1 is the ClCCl angle, and φ_2 is the HCH angle. Assuming that $\varphi_1 = \varphi_2 = 109°28'$, we find for $\mu_{CH_2Cl_2}$ a value of 1.63 D. The experimental value is 1.62 D (Stark effect).

TABLE 32. Influence of Induction Effects on the Dipole
Moments of the Bonds (in D) in Halogen-Substituted Methanes

Bond	CH_3Hal	CH_2Hal_2	$CHHal_3$	Bond	CH_3Hal	$CHHal_2$	CH_2Hal_3
H—C	0	0.23	0.32	H—C	0	0.13	0.16
C—F	1.81	1.45	1.22	C—Br	1.78	1.12	0.80
H—C	0	0.16	0.19	H—C	0	0.10	0.12
C—Cl	1.86	1.25	0.92	C—I	1.59	0.97	0.66

The method of calculating induced moments illustrated is
used mainly for compounds of the aliphatic and alicyclic series.
However, it can also be used to calculate the distribution of
charges and induced moments in aromatic and heteroorganic compounds [18, 19].

Investigation of the Influence of Induction Effects

on the Electronic Structure of Organic Compounds

by Means of Dipole Moments

Let us consider how the apparatus for determining the influence of induction effects on the overall dipole moments of organic
molecules is used to study features of their electronic structure
connected with these effects.

Dipole Moments of Halogen-Substituted
Methanes. Taking induction into account by Smith's method
leads to a value of the dipole moment of chloroform of 1.12 D,
which agrees well with the experimental value. This result can be
explained by a change in the magnitudes of the primary dipoles
C—H and C—Cl under the influence of the polarizing C—Cl groups.
Table 32 gives the values calculated from Smith's data [4] of the
values of the C—Hal and C—H moments in the halogen-substituted
methanes: $\mu_{C-Hal} = \varepsilon_{Hal}R_{C-Hal}$, $\mu_{C-H} = \varepsilon_H R_{C-H}$. These values
show that an accumulation of polar C—Hal bonds in the molecule
leads to the appearance of induced moments which considerably
reduce the effective moments of the C—Hal bonds. At the same
time, relatively small induced dipoles with the polarity H^+C^- appear in the C—H bonds.

TABLE 33. Calculated [11, 12] and Experimental [20] Values of the Dipole Moments of Halogen–Substituted Alkanes

Compound	μ_{calc}, D	μ_{exp}, D*	Compound	μ_{calc}, D	μ_{exp}, D*
CH_3F	1.81	1.79 (S)	CH_3CHBr_2	1.80	2.12 (B)
CH_2F_2	1.91	1.96 (S)	CH_3CHI_2	1.56	2.24 (G)
CHF_3	1.53	1.64 (S)	CH_3CF_3	2.39	2.32 (S)
CH_3Cl	1.86	1.87 (S)	CH_3CCl_3	1.71	1.77 (G)
CH_2Cl_2	1.63	1.62 (S)	CH_3CF_2Cl	2.19	2.14 (G)
$CHCl_3$	1.12	1.2 (S)	CF_3CF_2Cl	1.25	0.80 (G)
CH_3Br	1.78	1.80 (S)	CF_3CHF_2	1.66	1.54 (B)
CH_2Br_2	1.48	1.5 (S)	CCl_3CHCl_2	1.27	0.92 (S)
$CHBr_3$	0.98	1.02 (B)	CCl_3CH_2Cl	1.78	1.39[12]
CH_3I	1.59	1.65 (S)	$iso-C_3H_7Cl$	2.15	2.15 (G)
CH_2I_2	1.23	1.10 (B)	$iso-C_3H_7Br$	2.08	2.19 (G)
CHI_3	0.78	1.00 (G)	$iso-C_3H_7I$	2.01	1.99 (B)
$CFCl_3$	0.95	0.68 (G)	$(CH_3)_2CCl_2$	2.25	2.25 (G)
CF_2Cl_2	1.18	0.70 (G)	$tert-C_4H_9F$	2.11	1.96 (S)
C_2H_5F	1.95	1.96 (S)	$tert-C_4H_9Cl$	2.25	2.15 (S)
C_2H_5Cl	2.02	1.79 (S)	$tert-C_4H_9Br$	2.18	2.21 (S)
C_2H_5Br	1.95	2.01 (G)	$tert-C_4H_9I$	2.07	2.13 (S)
C_2H_5I	1.82	1.87 (G)	$CH_2—CH_2—CCl_2$ (ring)	1.49	1.58 (S) [21]
CH_3CHF_2	2.33	2.30 (S)	$CH_2—CH_2—CH_2CHBr$ (ring)	2.20	2.09 (S) [29]
CH_3CHCl_2	1.99	1.98 (B)	$CH_2—CH_2—CH_2—CH_2—CHBr$ (ring)	2.10	2.16 (B)
			$CH_2—CH_2—CH_2—CH_2—CHI$ (ring)	1.98	2.06 (B)

* In the selection of the experimental values, preference was given to the results obtained by measurements of the Stark effect in microwave spectra (S) and then to the results of measurements in the gas phase (G). In the absence of these data, the magnitudes of the dipole moments obtained in measurements on benzene solutions (B) are given.

It can be seen from the figures in Table 32 how important it is to take into account induction at small distances where two or more polar groups are attached to a single carbon atom. Table 33 gives the results of calculations of the dipole moments of a number of poly–halogen–substituted methanes, ethanes, and propanes published by Smith et al. [11, 12]. These results are in good agreement with the experimental figures [20].

Another important fact is that the direction of the resultant vector of the dipole moment predicted in calculations by Smith's method agrees well with experiments in those cases where the latter can be determined, such as those using microwave measurements [26].

Dipole Moments in Homologous Series. As follows from the calculations the results of which are given in Table 32, polar groups induce small dipole moments in C−H bonds. The magnitudes of the induced moments in C−H bonds fall rapidly with the distance of these bonds from the polar group. Thus, in ethyl chloride the moment of the C−H bond in the methylene group adjacent to the chlorine atom is 0.09 D, while the C−H bonds in the methyl group have a moment of only 0.02 D [12]. As calculations show, the moment induced in an alkyl radical rises in a homologous series of monosubstituted alkanes reaching its maximum at R = C_5H_{11}, C_6H_{13}. A further accumulation of methylene groups does not lead to an increase in the induced moment and is therefore not reflected in the overall dipole moment (Table 34).

It is interesting to note that calculations carried out by the LCAO MO method for the alkyl halides $C_nH_{2n+1}X$ also show an increase in the dipole moment up to n ≤ 6, after which the dipole moment ceases to depend on the length of the alkyl radical [22].

The dipole moments of compounds of a homologous series also depend on the degree of branching of the radical. It follows from the figures of Table 34 that the moments of isopropyl derivatives are always higher than those of n-propyl derivatives, and the moments of sec- and tert-butyl isomers are higher than those of the n-butyl isomers. This is due to the fact that in a branched radical the terminal C−C and C−H bonds are located closer to the polarizing group than the terminal bonds in a radical of normal structure. Consequently, the induced moment is greater in the branched radical. Another possible cause is some change in the polarizability of the C−H and C−X bonds in the branched isomers. This is shown by the small but quite definite dipole moment found by the microwave method for isobutane, which is 0.132 D [28].

Induction Effects of Substituents in the Aromatic Series. The figures given in Table 12 show that unlike those of the meta and para isomers the dipole moments of

TABLE 34. Dipole Moments (in D) of Homologous Series of Compounds R–X*

R	Cl		Br		I		NO_2		CN		HgBr
	gas	CCl_4	gas	CCl_4	gas	CCl_4	gas	B	gas	CCl_4	B
CH_3	1.83	1.86	1.82	1.73	1.70	1.48	3.50	3.10	3.94	3.38	2.42
C_2H_5	1.98	1.90	2.01	1.93	1.87	1.75	3.58	3.19	4.0	3.52	2.80
C_3H_7	2.04	1.93	2.15	1.97	2.01	1.79	3.72	—	4.05	—	3.28
iso–C_3H_7	2.15		2.19	—	—	—	3.73	—	—	3.62	—
C_4H_9	2.11	1.92	2.15	1.96	2.08	1.81	3.35	3.29	4.09	—	3.57
sec–C_4H_9	2.12	—	2.20	—	—	—	3.71	—	—	—	—
iso–C_4H_9	2.04	—	—	—	—	—	—	—	—	—	—
tert–C_4H_9	2.11	2.14	2.21	—	—	2.14	—	3.71	—	3.68	—
C_5H_{11}	2.12	1.94	2.13	1.96	—	1.85	—	—	—	—	3.47
C_6H_{13}	—	1.97	2.16	1.97	—	1.84	—	—	—	—	—
C_7H_{15}	—	—	2.15	1.97	—	1.79	—	—	—	—	—
C_8H_{17}	—	—	—	1.96	—	1.79	—	—	—	—	—
C_9H_{19}	—	—	—	1.94	—	1.82	—	—	—	—	—
$C_{10}H_{21}$	—	—	—	1.93	—	1.82	—	—	—	—	—
$C_{11}H_{23}$	—	—	—	—	—	1.83	—	—	—	—	—
$C_{12}H_{25}$	—	—	—	1.95	—	—	—	—	—	—	3.67
$C_{14}H_{29}$	—	—	—	1.90	—	—	—	—	—	—	—
$C_{16}H_{33}$	—	—	—	1.96	—	—	—	—	—	—	—
$C_{18}H_{37}$	—	—	—	1.95	—	—	—	—	—	—	—

* The values of the dipole moments given were determined in the gas phase (gas) or at 25°C in benzene (B) and carbon tetrachloride (CCl_4). Besides data from a handbook on dipole moments [20], the following sources have been used:

R – Cl . . . . . [23]   R – CN . . . . . . [25]
R – Br . . . . . [24,30] R – HgBr . . . . . [27]
R – I . . . . . [25]

ortho–disubstituted benzenes calculated by the usual vectorial scheme do not agree well with the experimental figures. It is natural to assume that the cause of the deviations found in the dipole moments may be the moments induced by each substituent in the neighboring substituent and also in the hydrocarbon part of the molecule. The difference between the experimental and calculated values of the dipole moments of a number of substituted naphthalenes, biphenyls, and anthracenes can be explained in the same way. Table 35 gives some data for the calculations of the dipole moments of dihalo–substituted aromatic compounds, taking into account the induction effect in accordance with formulas (V.1) and (V.3). It can be seen that the inclusion of induced dipoles in the

TABLE 35. Dipole Moments of Di-Halogen-Substituted Aromatic Compounds

Compound	μ_{exp}, D [20]	μ_{calc} without taking the induction factor into account	μ_{calc} taking the induction factor into account	Literature
o-Dichlorobenzene	2.27	2.76	2.30	[2]
o-Dibromobenzene	2.26	2.70	2.09	[2]
o-Diiodobenzene	1.69	2.43	1.50	[2]
o-Bromoiodobenzene	1.86	2.58	2.09	[2]
m-Dichlorobenzene	1.48	1.59	1.48	[2]
m-Dibromobenzene	1.46	1.57	1.43	[2]
2,3-Dichloronaphthalene	2.55	2.98	2.48	[31]
1,8-Dichloronaphthalene	2.82	3.12	2.80	[6]
1,8-Dichloroanthracene	3.2	3.12	3.06	[6]

calculated magnitude of the dipole moment substantially improves agreement with the experimental value.

This improvement is particularly marked if the substituents in the aromatic ring are located adjacent to one another, as in the ortho derivatives of benzene and in 2,3-dichloronaphthalene. The difference between the moments calculated with and without taking induction into account is evened out as the substituents move away from one another. It is less important in the case of 1,8-dichloro-naphthalene and is very small for meta-disubstituted benzenes and 1,8-dichloroanthracene.*

More detailed compilations of the dipole moments calculated taking induction into account for benzene compounds [2, 32], biphenyl compounds [33], and naphthalene compounds [31, 34] can be found in the respective literature mentioned.

---

* In an analysis of the figures of Table 35, it must also be borne in mind that if, in accordance with the results of electron-diffraction studies [35], it is assumed that the halogen atoms in o-dihalo-substituted derivatives depart from the plane of the ring by some small angle $\varphi \simeq 10°$, the dipole moment calculated without taking induction into account agrees fairly well with the experimental value.

2. Conjugation of an Unsaturated System with Polar Groups. Mesomeric Moments. Interaction Moments.

The values of the group moments given in Table 10 show that similarly substituted aliphatic and benzene derivatives have substantially different values of the dipole moment. The reason for this cannot be only the difference in the bond moments caused by the different hybridizations of the carbon atoms. In actual fact, the dipole moments of monosubstituted benzenes differ from those of the analogous derivatives of other aromatic compounds.

CH$_3$—X

X = NO$_2$, μ = 3.10 D
X = NH$_2$, μ = 1.46 D

X = NO$_2$, μ = 4.01 D
X = NH$_2$, μ = 1.53 D

X = NO$_2$, μ = 4.36 D
X = NH$_2$, μ = 1.74 D

X = NO$_2$, μ = 3.98 D
X = NH$_2$, μ = 1.49 D

X = NO$_2$, μ = 4.36 D
X = NH$_2$, μ = 1.77 D

—CH=CH—

X = NO$_2$, μ = 4.56 D
X = NH$_2$, μ = 2.07 D

It is clear that the variations in the dipole moments of these aromatic derivatives are a consequence of the dissimilar conjugation of the substituent with different aromatic nuclei, i.e., dissimilar mesomeric and π-induction effects. These two effects are completely absent in saturated systems. It is therefore understandable that the first attempts to determine their influence on dipole moments were associated with comparisons of the moments of aliphatic and aromatic compounds and in the first place, with compounds of the benzene series, for which the largest amount of data has been accumulated.

Mesomeric Effects of Substituents in the Benzene Series

The mesomeric interaction of a substituent with the benzene nucleus must correspond to the appearance of an additional moment directed from the substituent to the nucleus for +M substituents and in the opposite direction for −M substituents. Since the polarity of the C−X bond is generally such that the carbon atom acts as the positive pole (Table 8), the mesomeric interaction leads to a reduction in the dipole moment of a substituted benzene

TABLE 36. Mesomeric Moments
Calculated from Formula (V.17) [38, 47]

Group	μ_M, D	Group	μ_M, D
—CH$_3$	0.35	—SCH$_3$	0.44
$>$O	0.15	—CCl$_3$	−0.5
		—CF$_3$	−0.2
—NH$_2$	1.02	—COCH$_3$	−0.46
—N(CH$_3$)$_2$	1.66		
—F	0.41	$>$C=O	−0.3
—Cl	0.41		
—Br	0.43	—C≡N	−0.45
—I	0.50	—NO$_2$	−0.76
—OCH$_3$	0.96	$>$SO$_2$	−0.6
—SH	0.8	Si(CH$_3$)$_3$	0.42
		—SiCl$_3$·	−0.37

as compared with a substituted alkane in the case of +M substituents and an increase in the case of −M substituents.

Sutton [36] proposed to call the difference in the values of the dipole moments of the corresponding saturated and benzene derivatives the mesomeric moment μ_M. For irregular groups the moments must be subtracted vectorially. Here it is assumed that the mesomeric moment arises only in the direction of the axis connecting the first atom of a substituent to the benzene nucleus, and the appearance of $\vec{\mu}_M$ is not reflected in the magnitudes of the components of the dipole moment not collinear with it [37, 38].

$$\vec{\mu}_M = \vec{\mu}_{CH_3X} - \vec{\mu}_{PhX} \qquad (V.17)$$

Values of the mesomeric moments calculated according to (V.17) are given in Table 36. The sign and magnitude of $\vec{\mu}_M$ correspond to the direction and relative force of the mesomeric action of the substituent. The values given in Table 36 correspond in the majority of cases to ideas on the mesomeric effects of the substituents based on data relating to their reactivities [39]. Thus, correlations have been established between the mesomeric moments defined by equation (V.17) and the substituent constants of the σ_c^0 [40] and σ_c [41] (for a review, see the book by Zhdanov and Minkin [1]).

The success of these correlations is due to the fact that the mesomeric moments calculated according to (V.17) reflect correctly the relative contributions of the conjugation effects to the dipole moments in many cases. However, the absolute values of these effects expressed in terms of dipole moments do not correspond to Sutton's mesomeric moments. In actual fact, the difference in the dipole moments of a substituted methane and a substituted benzene is composed of the following individual moments [37, 42, 43]:

$$\vec{\mu}_{CH_3X} - \vec{\mu}_{PhX} = \vec{\mu}_{I\pi} + \vec{\mu}_M + \Delta\vec{\mu}_i + \Delta\mu_{C-H} + \Delta\vec{\mu}_{C-X} \qquad (V.18)$$

where $\vec{\mu}_{I\pi}$ and $\vec{\mu}_M$ are the moments due, respectively, to the π-induction effect [1, 48] and the true mesomeric effect; $\Delta\vec{\mu}_i$ is the difference in the moments induced by a polar group in the methyl and phenyl nuclei; and $\Delta\vec{\mu}_{C-H}^-$ and $\Delta\vec{\mu}_{C-X}$ are the differences in the moments of the corresponding bonds caused by the dissimilar hybridizations of the carbon atom in the aliphatic and aromatic series.

Another attempt has been made to take the individual members in equation (V.18) into account. It has been shown, for example, that it is more correct to compare the dipole moments of benzene derivatives not with the methyl but with the corresponding tert-butyl compounds, since the moments induced by polar groups in the phenyl and the tert-butyl radicals are approximately the same [44, 45]. Values of the mesomeric moments can be refined if the term $\Delta\mu_{C-H}$ is taken into account [43]. However, even if it were possible to evaluate all the last three terms in equation (V.18) correctly by comparing dipole moments of saturated and unsaturated compounds, it is in principle impossible to separate the value of $\mu_{I\pi}$ from $\vec{\mu}_M$.

In view of this, another approach to the evaluation of mesomeric moments has been proposed [37, 42, 46]. It is based on a comparison of the moments of monosubstituted compounds of, for example, nitrobenzene (I) and its mesityl (II) or duryl (III) derivatives.

I   II   III

$\mu = 4.01\ D$ $\mu = 3.67\ D$ $\mu = 3.62\ D$

In view of their centrosymmetrical arrangement, the methyl groups in compounds (II) and (III) do not cause an additional moment through their inherent group moments; however, for spatial reasons the nitro group in (II) and (III) cannot be arranged in the plane of the benzene ring and deviates from it by some angle φ approximating to 90°.* This obviously leads to the suppression of conjugation of the substituent with the nucleus and a levelling out of the mesomeric moment. At the same time, all the other components of equation (V.18) keep the same values in (II) and (III) as in (I). Thus, the difference in the values of the dipole moments of (II) or (III) and (I) is a measure of the true mesomeric moment of the nitro group:

$$\vec{\mu}_M = \vec{\mu}_{ArX} - \vec{\mu}_{PhX} \tag{V.19}$$

where Ar is a mesityl or duryl radical.

Like (V.18), relation (V.19) assumes that in the case of irregular groups only the components of the overall moment acting in the direction of the bond between the substituent and the nucleus are considered. The method of calculation is clear from the example of the calculation of the mesomeric moment of the dimethyl-amino group [50].

IV   V

The dipole moment of dimethylaniline (IV) (1.61 D) can be resolved into two components: parallel and perpendicular to the N—Ph axis (see Table 10). Assuming, then, that the perpendicular component in dimethylmesidine (V) has the same magnitude as in

---

* According to X-ray data [49], the angle φ between the planes of the benzene nucleus and the nitro group in compound (II) is 66°.

dimethylaniline (IV), we carry out an analogous operation with the overall moment of (V) (1.03 D). In this, the direction of the parallel component of the vector of the overall moment in (V) is in the opposite direction to that in (IV), since the value of the dipole moment of (V) is higher than that of dimethyl-2,6-xylidine (0.94 D). Hence, according to (V.19)

$$\mu_M = 0.66 - (-1.40) = 2.06\ D$$

The determination of the mesomeric moment by comparing the dipole moments of compounds containing the mesomeric group and its p-methyl-substituted derivative gives very important information concerning the features of the electronic interaction of a substituent with an aryl nucleus in many cases. Thus, from a comparison of the moments of compounds of types (VI) and (VII) [65]

$$R_2P-\!\!\!\!\bigcirc \qquad\qquad R_2P-\!\!\!\!\bigcirc\!\!-CH_3$$

VI VII

R = H, $\mu = 1.11\ D$ R = H, $\mu = 1.41\ D$

R = C_4H_9, $\mu = 1.45\ D$ R = C_4H_9, $\mu = 1.55\ D$

it follows that the mesomeric moment of phosphino and dialkyl-phosphino groups coincide in direction with the moment of the methyl group in (VII), i.e., are directed from the ring to the phosphorous atom. Consequently, the p,π-conjugation effect, like that which is characteristic for the nitrogen analogs (VIII) and (IX) of compounds (VI) and (VII) and must lead to a reversed direction of the mesomeric moment, is suppressed by d,π-conjugation due to the electron-accepting properties of the PR_2 group.

$$R_2N-\!\!\!\!\bigcirc \qquad\qquad R_2N-\!\!\!\!\bigcirc\!\!-CH_3$$

VIII IX

R = H, $\mu = 1.53\ D$ R = H, $\mu = 1.32\ D$

R = C_4H_9, $\mu = 1.91\ D$ R = C_4H_9, $\mu = 1.57\ D$

The introduction of voluminous groups into the ortho position of a benzene nucleus inhibits the conjugation of a substituent with the nucleus only in those cases where the effective volume of the substituent is sufficiently large to make a planar configuration of the molecule impossible and to turn the substituent relative to the plane of the nucleus by a sufficiently large angle φ. If, however, the volume of the substituent (for example, OH, Cl, Br) is small,

TABLE 37. Dipole Moments of Compounds of the Type ArX (Benzene, 25°C)

X	Ar	μ_{exp}, D [20]	φ *
NO_2	C_6H_5	4.01	
	$C_6H_2(CH_3)_3$-2,4,6	3.67	66°, X-ray [49]
	$C_6H(CH_3)_4$-2,3,5,6	3.62	
	$C_6H(uзo\text{-}C_3H_7)_4$-2,3,5,6	3.57 [51]	
	$C_6H_2Br_3$-2,4,6	3.17	
$N(CH_3)_2$	C_6H_5	1.61	
	$C_6H_2(CH_3)_3$-2,4,6	1.03	
	$C_6H_2Br_3$-2,4,6	1.05	
$NHCH_3$	C_6H_5	1.68	
	$C_6H_2(CH_3)_3$-2,4,6	1.22 [50]	
	$C_6H_2Br_3$-2,4,6	1.68	
NH_2	C_6H_5	1.53	
	$C_6H_2(CH_3)_3$-2,4,6	1.45	
	$C_6H(CH_3)_4$-2,3,5,6	1.45	
	$C_6H_2Br_3$-2,4,6	1.69	
OH	C_6H_5	1.55	
	$C_6H_2(CH_3)_3$-2,4,6	1.43	
	$C_6H_2(mpem\text{-}C_4H_9)_3$-2,4,6	1.55	12°, Kerr [52]
	$C_6H(CH_3)_4$-2,3,5,6	1.68	
	$C_6H_2Br_3$-2,4,6	1.55	
OCH_3	C_6H_5	1.28	
	$C_6H_2(CH_3)_3$-2,4,6	1.26 [52]	90°, Kerr [52]
	$C_6H_2Br_3$-2,4,6	1.39	
CHO	C_6H_5	2.96	
	$C_6H_2(CH_3)_3$-2,4,6	2.95	0°, Kerr [52]
$COCH_3$	C_6H_5	2.96	
	$C_6H_2(CH_3)_3$-2,4,6	2.81	90°, Kerr [52]; 80°, UV [53]
	$C_6H(CH_3)_4$-2,3,5,6	2.81	90°, Kerr [52]
Br	C_6H_5	1.57	
	$C_6H_2(CH_3)_3$-2,4,6	1.52	
	$C_6H(CH_3)_4$-2,3,5,6	1.55	
SCH_3	C_6H_5	1.31 [54]	
	$C_6H(CH_3)_4$-2,3,5,6	1.50 [54]	

* Symbols for the method of determination: X-ray—the X-ray method; Kerr—molecular polarizability, Kerr effect; UV—electronic absorption spectra.

it does not experience a large steric interaction due to ortho substituents in the nucleus. The substituting group is located in the plane of the nucleus or deviates only slightly from it. Consequently, the dipole moments of 2,4,6-trimethyl or 2,3,5,6-tetramethyl derivatives of phenol and the halobenzenes have practically the same moments as the unsubstituted phenol and halobenzenes [20].

TABLE 38. True Mesomeric Moments of Polar Groups in the Benzene Nucleus

Substituent	μ_M, D	μ_M, D(SCF MO)	Substituent	μ_M, D	μ_M, D(SCF MO)
NH_2	$+1.67$ [50]	1.72 [55]	OCH_3	$+0.9$ [45]	0.5 [56]
$NHCH_3$	$+1.93$ [50]	—	$COCH_3$	-0.23	—
$N(CH_3)_2$	$+2.06$ [50]	2.64 [58]	NO_2	$-0.44*$	0.8 †

* Calculated as the difference in the dipole moment of nitrobenzene and its 2,3,5,6-tetrapropyl derivative.

† Approximate evaluation from the contribution ($\simeq 6\%$ [57]) to the ground state of the electronic configuration with charged transfer.

Table 37 gives data on the dipole moments of various symmetrically substituted benzene derivatives containing voluminous substituents in both ortho positions with respect to the polar group. The figures given in Table 37 show that steric inhibition of the mesomeric effect is greatest for voluminous polar groups. Thus, the dipole moment of mesidine differs from the dipole moment of aniline far less than the dipole moment of dimethylmesidine differs from the moment of dimethylaniline. The introduction of ortho-methyl groups into the molecule of benzaldehyde does not deflect the aldehyde group from the plane of the benzene nucleus, but an acetyl group is deflected by ortho-methyl groups by 90°.

Table 38 gives the values of the mesomeric moment calculated on the basis of the data of Table 37 from equation (V.19). These values are compared with the values calculated by the SCF MO method. The values given in Table 38 for the mesomeric moments show the comparatively low mesomeric effects of electron-accepting groups. This is in harmony with recently developed ideas on the low degree of conjugation of these groups (in the absence of electron-donating groups) with the aromatic nucleus [49, 59].

Polar Conjugation of Substituents through a p-Phenylene System

If there is an electron-accepting group in the p (or o) position to an electron-donating substituent in the benzene nucleus, the polar conjugation effect is observed, which consists in the displacement of electrons from the electron-donating to the electron-

TABLE 39. Dipole Moments of Compounds of the

Type D —⟨ ⟩—A in Benzene at 25°C (in D) [20]

D represents an electron-donating group and A an electron-accepting group.

Compound		μ_{exp}	μ_{PhD}	μ_{PhA}	μ_{calc}	μ_{int}
D	A					
$N(CH_3)_2$	$N{=}O$	6.90	1.61	3.09	4.41	2.63
	CHO	5.60		2.96	4.16	1.53
	NO_2	6.93		4.01	5.46	1.48
	$COCH_3$	5.05		2.96	4.00	1.20
	CF_3	4.62		2.54	4.01	0.61
	$CH{=}CH_2$	2.17		0.12	1.66	0.50
	CN	5.90		4.05	5.50	0.41
	Cl	3.29		1.59	3.10	0.20
NH_2	SO_2CF_3	6.88 [61]	1.53	4.32	5.44	1.48
	NO_2	6.33		4.01	5.16	1.19
	SF_5	5.31		3.44	4.60	0.73
	$COCH_3$	4.43		2.96	3.89	0.67
	Cl	2.99		1.59	2.85	0.15
OH	CHO	4.23	1.55	2.96	3.34	1.05
	NO_2	5.07		4.01	4.30	0.82
	SO_2CH_3	5.32		4.73	4.97	0.71
	CN	4.95		4.05	4.34	0.63
	Cl	2.27		1.59	2.22	0.06
Br	NO_2	2.66	1.57	4.01	2.44	0.22
	CN	2.64		4.05	2.48	0.16

accepting substituents through the phenylene system. This displacement of electrons leads to the appearance of an additional moment the vector of which coincides in direction with the vectors of the mesomeric moments of the substituents. Consequently, the observed value of the dipole moments of molecules of the type under consideration will exceed the values calculated from the sum of the moments of the polar groups. It has been proposed to call the vectorial difference between the experimental and calculated dipole moments the interaction moment [60]:

$$\vec{\mu}_{int} = \vec{\mu}_{exp} - \vec{\mu}_{calc} \qquad (V.20)$$

Table 39 gives the experimental values for a series of p-disubstituted benzenes each of which contains one electron-donating and one electron-accepting group, compared with those calculated by the additive scheme according to formula (III.10). In the last column the interaction moments calculated from (V.20) are given. In practice, the calculation reduces to the solution of the quadratic equation

$$\mu^2_{int} + 2\mu_{int}\left(\mu_D \cos\theta_D + \mu_A \cos\theta_A\right) + \mu^2_{calc} - \mu^2_{exp} = 0 \qquad (V.21)$$

where μ_D and μ_A are the group moments of the electron–donating and electron–accepting groups and θ_D and θ_A are the angles that the vector of the group moment makes with the axis joining the first and fourth carbon atoms of the benzene nucleus. In accordance with the definition, of the two values of μ_{int} obtained by solving equation (V.21) the positive one is taken.

The order of magnitude of the interaction moments in each of the reaction series given in Table 39 with a fixed electron–donating and a varying electron–accepting group are in agreement with developing ideas on the relative strengths of the accepting action of the latter. The σ_c^- constant of an electron–accepting substituent is a measure of polar coordination in compounds of this type [1]. If it is assumed that the distance between the centers of the positive and negative charges induced by the polar conjugation effect is approximately the same for all p–disubstituted benzenes, the following relation may be expected to be satisfied:

$$\mu_{int} = \varkappa\sigma_c^- \qquad (V.22)$$

where $\varkappa$ is a proportionality constant characterizing the capacity of the electron–donating group for conjugation with electron–accepting substituents.

In actual fact, as is shown in Fig. 24 for p–substituted dimethylanilines, relation (V.22) is satisfied fairly well. It is obvious that if a reaction series is formed by compounds with a fixed electron–accepting and a varying electron–donating group, σ_c^- in equation (V.22) must be replaced by the σ_c^+ constants of the electron–donating groups, and the factor $\varkappa$ will characterize the conjugation properties of the electron–accepting group.

Some other examples of the correlation of the interaction moments with the substituent constants can be found in Zhdanov and Minkin's monograph [1].

As can be seen from the data of Table 39, the interaction moment makes an extremely substantial contribution to the total moment for many of the compounds given in it. The magnitude of

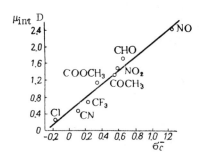

Fig. 24. Correlation of the dipole moments of p-substituted dimethyl-anilines with the σ_c^- constants of the substituents [1].

this moment can also be judged by comparing dipole moments in reaction series of the type

D
│
◇
‖
│
A

I

H₃C D CH₃
   ◇
H₃C   CH₃
     │
     A

II

If the electron–donating group (D) and the electron–accepting group (A) have sufficiently large volumes and are displaced by the o–methyl groups from the plane of the benzene nucleus by an angle $\varphi = 90°$, the dipole moment of the durylene derivative (II) can be calculated from the moment of the phenylene derivative (I) in the following way:

$$\vec{\mu}_{II} = \vec{\mu}_I - \vec{\mu}_{M,D} - \vec{\mu}_{M,\ A} - \vec{\mu}_{int} \qquad (V.23)$$

Unfortunately, at the present time experimental data on dipole moments are available for only a few compounds of type (II). Moreover, in all cases the acceptor substituent is a nitro group which, as mentioned above, is not displaced completely from conjugation with the nucleus and the donor substituent. Nevertheless, the values of the dipole moments of compounds (II) are substantially lower than those of (I) (Table 40). In the last column are given the value of the interaction moments in the durylene system calculated for those compounds for which it has been possible to determine the vector of the group moment of the donor group in monosubstituted durene.

The observation of an interaction moment of a substituting group with p-substituents the electron–donating or electron–accepting nature of which has been established can give valuable information on the nature of electronic conjugation effects. To illustrate this point, let us consider a few examples from the field of heteroorganic compounds.

A sulfur atom may take part in a dual type of electronic interaction with the π-system of an aromatic nucleus. Most typical for sulfur is the electron–donating effect due to the inclusion of

TABLE 40. Dipole Moments (in D) of
Compounds of Type (I) and (II) (Ar–NO$_2$) in
Benzene at 25°C

D	μ_I	μ_{II}	μ_{int}
NMe$_2$	6.93	4.11	—
NH$_2$	6.33	4.98	0.30
OH	5.07	4.08	—
OC$_2$H$_5$	4.86*	3.69	—
Br	2.66	2.36	0.29

* Value for p-nitroanisole.

the occupied 3p orbitals of sulfur in the general π-system. The consequence of this effect is the appearance of a considerable interaction moment in p-nitrothioanisole (III).

$\mu_{exp} = 4.43$ [54]; $\mu_{calc} = 3.99\,D$

Within the framework of the method of valence bonds, such an electronic interaction of the methylthio and nitro groups through a p-phenylene system corresponds to some contribution of the polar structure (IIIa) to the ground electronic state of the molecule.

The methylthio group, however, may also exhibit electron-accepting properties through the vacant 3d orbitals of the sulfur atom (d, π conjugation). This effect is assumed in order to explain a number of chemical reactions and some features of the electronic spectra of organosulfur compounds [62]. In actual fact, the experimental value of the dipole moment of p-dimethylaminothioanisole (IV) is less than the calculated value and the vector of the interaction moment in (IV) is opposite to that in (III).

$\mu_{exp} = 2.82$ [54]; $\mu_{calc} = 2.24\,D$

$\mu_{exp} = 1.76$ [54]; $\mu_{calc} = 1.84\,D$

The cause of the appearance of an interaction moment in (IV) can only be d,π conjugation with the participation of the vacant d-orbitals of the sulfur corresponding to a contribution of the polar structure (IVa). An additional confirmation is the absence of an interaction moment in the molecule of p–dimethylaminoanisidine (V). Even higher interaction moments, showing d,π conjugation, are found for p–substituted aryl sulfoxides, aryl sulfones [61], and aryl sulfonyl fluorides [63]. It is stated [61] that the cause of this is the appearance of an effective positive charge on the sulfur in the functional groups mentioned which favors the transfer of elec-trons into the vacant d-orbitals of the sulfur.

The appearance of an interaction moment due to d,π–conjuga-tion is characteristic not only for compounds with sulfur–containing functional groups but also for all other benzene derivatives having substituents with low–energy d–orbitals, for example, compounds (VI) and (VII).

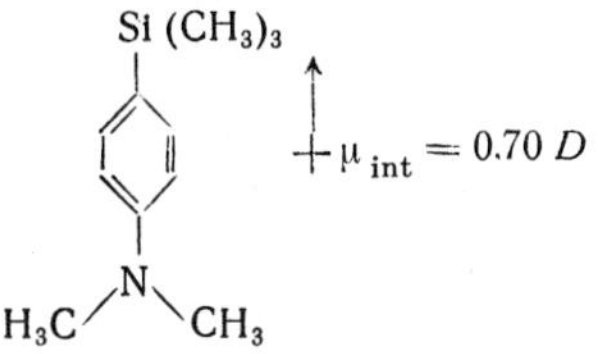

$\mu_{int} = 0.70\ D$

VI

$\mu_{exp} = 1.84\ [64];\ \mu_{calc} = 1.24\ D$

$\mu_{int} = 0.77\ D$

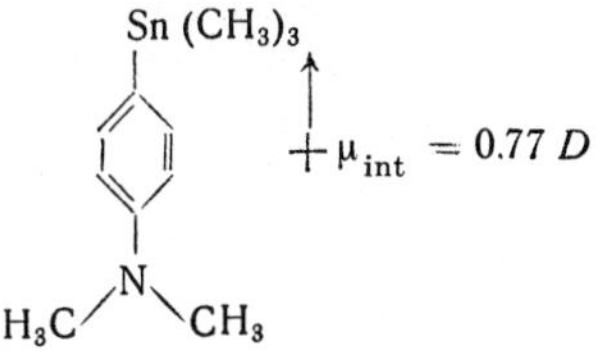

VII

$\mu_{exp} = 1.83\ [64];\ \mu_{calc} = 1.19\ D$

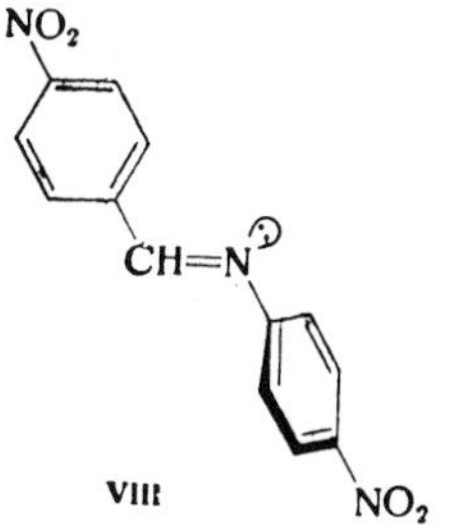

VIII

$\mu_{exp} = 3.56\ [66];\ \mu_{calc} = 1.6\ D$

In some cases the existence of an interaction moment may also give an idea of the geometrical structure of the molecule of

TABLE 41. Dipole Moments of Substituted Biphenyls, Stilbenes, and 1,4-Diphenylbutadienes in Comparison with the Dipole Moments of Benzene Derivatives [20]

Compound	μ, D	μ of corresponding benzene derivative, D
Biphenyl		
4-Fluoro-	1.49	1.47
4-Chloro-	1.66	1.59
4-Bromo-	1.66	1.57
4-Iodo-	1.54	1.40
4-Nitro-	4.36	4.01
4-Cyano-	4.33	4.05
4-Hydroxy-	2.21	1.55
4-Amino-	1.74	1.53
4-Dimethylamino-	2.04	1.61
4-Amino-4'-bromo-	3.30	3.01
4-Amino-4'-nitro-	6.42	6.33
4-Dimethylamino-4'-nitro-	6.93	6.93
Stilbene		
4-Bromo-	1.67	1.57
4-Nitro-	4.56	4.01
4-Cyano-	4.42	4.05
4-Amino-	2.07	1.53
4-Dimethylamino-	2.41	1.61
4-Amino-4'-bromo-	3.56	3.01
4-Amino-4'-nitro-	6.83	6.33
4-Dimethylamino-4'-nitro-	7.42	6.93
4-Amino-4'-sulfonamido-	6.71	6.22
Diphenylacetylene [70]		
4-Methyl	0.57	0.37
4-Chloro-	1.68	1.59
1,4-Diphenyl-1,3-butadiene		
4-Nitro-	4.75	4.01
4-Dimethylamino-	2.58	1.61

the compound under investigation. Thus, the dipole moment of
p,p'-dinitrobenzylideneaniline (VIII) is considerably higher than
that of benzylideneaniline and the p,p'-dimethyl- and -dichloro-
substituted derivatives, in spite of the fact that the trans structure
has been reliably established for all these compounds. The reason
is the nonplanar structure of the molecule of the aromatic azomethines [67, 69] which ensures the possibility of the conjugation
of the unshared pair of electrons of the amine nitrogen with the π-
electrons of the nitro group through the p-phenylene system.

Aromatic Compounds with Several Uncondensed Rings

As has been shown above with individual examples, the dipole
moments of substituted biphenyls and stilbenes are higher than
those for the corresponding substituted benzenes. Fuller data confirming this statement are given in Table 41.

The fact that the moments of the substituted biphenyls, stilbenes, tolanes, and diphenylbutadienes are larger than the moments
of the analogous benzene compounds cannot be ascribed only to the
induction moment caused by the polar groups in the neighboring
aryl nuclei. In actual fact, the dipole moments of 3-substituted biphenyls and stilbenes differ from those of the benzene derivatives
by not more than 0.1-0.2 D. The reason for the higher values of
the dipole moments of the uncondensed aromatic compounds is,
therefore, the larger mesomeric moments and interaction moments
than in the benzene derivatives.

It may be assumed that the increase in the mesomeric interactions and polar conjugations is connected with an increased conjugation of the substituents with the aromatic system, leading to
the situation that electron-donating substituents donate and electron-accepting substituents accept a greater proportion of an electron than in the benzene series.

A basis for this assumption is formed by the results of determinations [70] of the mesomeric effects of substituents of several
series of compounds with structures of the type $R-\langle\!\!\!\!\!\bigcirc\!\!\!\!\!\rangle-M-X$,
where M is a bridging group and X is an aromatic radical, the ef-
fective electronegativity of which varies as a result of the directed
changes in structure shown in Table 42.

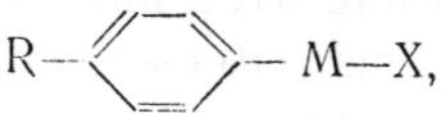

TABLE 42. Mesomeric Moments of Substituents in a Series
of Uncondensed Aromatic Compounds [70]

Type of compound	Substituent	μ_{exp}, D	μ_M, D
R—⟨ ⟩—⟨ ⟩	$N(CH_3)_2$	2.04	2.17
	NH_2	1.83	1.32
	Cl	1.66	0.35
	NO_2	4.36	—1.16
R—⟨ ⟩—⟨ ⟩N	NH_2	4.09	1.40
	H	2.51	—
	NO_2	1.90	—1.16
R—⟨ ⟩—⟨ ⟩N→O	H	4.52	—
	NO_2	0	—1.17
R—⟨ ⟩—CH=CH—⟨ ⟩	$N(CH_3)_2$	2.41	2.57
	NH_2	2.07	1.65
	CH_3	0.34	0.34
	OCH_3	1.13	0.91
	NO_2	4.56	—1.31
R—⟨ ⟩—CH=CH—⟨ ⟩N	OCH_3	3.55	1.34
	CH_3	3.19	0.49
	H	2.70	—
	Cl	1.20	0.51
	NO_2	1.85	—1.17
R—⟨ ⟩—CH=CH—⟨ ⟩N→O	OCH_3	5.56	1.61
	CH_3	4.86	0.37
	H	4.49	—
	NO_2	0	—1.24
R—⟨ ⟩—C≡C—⟨ ⟩	CH_3	0.57	0.57
	Cl	1.68	0.37
R—⟨ ⟩—C≡C—⟨ ⟩N	OCH_3	3.90	1.62
	CH_3	3.25	0.46
	H	2.79	—
	Cl	1.23	0.45
	NO_2	1.63	—1.17
R—⟨ ⟩—C≡C—⟨ ⟩N→O	H	4.29	—
	Cl	3.13	0.85

The mesomeric effects of substituents in compounds of this type
are calculated by means of an expression similar to (V.19):

$$\mu_M = \mu_{exp} - \mu_{PhMX} - \mu_{AlkR} \qquad (V.24)$$

A consideration of the data of Table 42 and a comparison of
them with those of Table 36 shows that 1) the mesomeric effect of
a substituent rises with an increase in the extent of the conjugated
system, i.e., in the sequence benzene—biphenyl—stilbene, and 2)
the mesomeric moment of a +M substituent rises with an increase
in the acceptor properties of the nucleus X, i.e., in the sequence
phenyl—4-pyridyl—4-pyridyl N-oxide.

On the other hand, according to Nazarova [71] and Vasil'eva [72], the main reason for the increase in dipole moments of derivatives of biphenyl, stilbene, etc., as compared with the corresponding benzene derivatives consists in the increase in the distance between the centers of gravity of the positive and negative charges and the larger extent of the aromatic systems, while the actual magnitudes of the charges on the electron-donating and electron-accepting groups undergo no appreciable changes.

The deduction of the limited donor and acceptor capacity of substituents was, however, made on the basis of very rough assumptions concerning the position of the centers of localization of the positive and negative poles of the molecular dipole and is not confirmed by the results of rigorous molecular-orbital calculations [58]. For a clearer illustration, let us consider the simplest model in which the electron-donating substituents are modelled by the carbanion CH_2^- and electron-accepting substituents by the carbocation CH_2^+. This model excludes the influence of the primary induction effects and to a certain approximation takes into account pure mesomeric effects. The molecular diagrams (I)–(V) show the distribution of the charges* calculated by Longuet-Higgins' method [73]:

---

* The values of the charges on the atoms must be taken with a positive sign for the $-CH_2^+$ derivatives and with a negative sign for the $-CH_2^-$ derivatives.

It can be seen clearly from the figures given that the donor (acceptor) capacity of a substituent rises appreciably with an increase in the extent of the conjugated system. At the same time, an accumulation of charges in the remote parts of the conjugated system must lead in real molecules to an increase in the distance between the centers of the negative and positive charges. Thus, it may be assumed that it is in fact the combination of the two factors mentioned that leads to the observed rise in the dipole moments of biphenyls, stilbenes, tolanes, and 1,4-diphenylbutadienes as compared with benzene derivatives.

Of course, the deduction made does not lead to the conclusion that the assumption of the limited donor and acceptor capacity of the substituents has no force. For example, calculation by the scheme given leads, for the planar polyphenylenes* with an arbitrary number n of nuclei, to the following expression of the charge of a CH_2 group:

$$q_{CH_2} = 2^{2n} \left[2^{2n} + 3 \sum_{m=0}^{m=n-1} 2^{2m} \right]^{-1}$$

It is easy to show that $\lim_{n \to \infty} q_{CH_2} = 0.5$, while already in the case of quaterphenyl (n = 4), $q_{CH_2} = 0.5002$. Consequently, there is a limiting value of the charge that a substituent conjugated with an aromatic system can acquire, but this limit depends on the nature of this system and has a different value for each homologous series.

Aromatic Compounds with Condensed Nuclei: Naphthalene

Derivatives

Among the derivatives of aromatic compounds with condensed nuclei, compounds of the naphthalene series have been studied to the greatest extent.

The dipole moments of monosubstituted naphthalenes are subject to a strict empirical law [34, 74-76]: β-substituted compounds are more polar than α-substituted ones. This law is confirmed by the data of Table 43.

---

* For the α,ω-diphenylpolyenes, the alternation of the lengths of the bonds in the polyene chain must be taken into account.

TABLE 43. Dipole Moments of α- and β- Monosubstituted
Naphthalenes in Benzene (in D) [20, 76]

Substituent	Position of the substituent		Substituent	Position of the substituent	
	α	β		α	β
CH_3	0.37	0.44	OH	1.45	1.54
C_4H_9	0.68	0.75	OCH_3	1.26	1.28
$COCH_3$	3.03	3.18	$OCOCH_3$	1.70	1.86
COOH	1.85 *	1,95 *	NH_2	1.49	1.77
$COOCH_3$	1.93	2.03	NO_2	3.98	4.36
CN	4.16	4.36	$NHCOCH_3$	3.46	3.79
F	1.42	1.49	N_3	1.36	1.60
Cl	1.56	1.72	NCO	2,30	2.34
Br	1,58	1.71	SO_2NH_2	5.12 *	5.27 *
I	1.43	1.56			

* In dioxane.

A comparison of the figures of Table 43 with the dipole moments of monosubstituted benzenes (Table 10) also leads to the
conclusion that the latter are approximately equal to the moments
of the α-substituted naphthalenes, while the moments of the β-
substituted naphthalenes are somewhat higher. In some cases,
this fact may be connected with the peri effect: the spatial interaction of the peri-hydrogen atom with an α-substituent, which
leads to a departure of the α-substituent from the plane of the
naphthalene nucleus and to the partial suppression of conjugation.
In actual fact, the α-nitro group in 1,5-dinitronaphthalene is displaced from the plane of the nucleus by an angle of 49° [77]. As a
result, the mesomeric moments of voluminous substituents present in the α position must be lower than for β substituents.

However, such an explanation loses its force in the case of
those substituents that are not deflected from the plane of the nucleus by a peri hydrogen atom (for example, OH, F, and other
halogens). The mesomeric effect of a substituent in the α position
is greater than in the β position. This follows, for example, from
a consideration of the following alternant cations (or anions):

TABLE 44. Dipole Moments of Nitronaphthylamines in Benzene at 25°C [20, 81]

Position of substituent		μ_{exp}, D	μ_{calc}, D	μ_{int}, D	Position of substituent		μ_{exp}, D	μ_{calc}, D	μ_{int}, D
NO_2	NH_2				NO_2	NH_2			
3	1	5.14	5.05	0.09	4	2	4.70	4.84	—0.15
4	1	6.40	5.07	1.35	5	2	5.03	4.84	0.21
5	1	5.17	5.07	0.09	6	2	6.04	5.69	0.36
6	1	5.14	5.05	0.09	7	2	5.89	5.22	0.72
7	1	4.41	4.11	0.43	8	2	4.47	3.76	0.95

Rigorous molecular-orbital calculations of the electronic structure of the α- and β-naphthols [78] and of the α- and β-nitronaphthalenes [79] also confirm this conclusion. The cause of the increase in the dipole moment of a β-substituted naphthalene as compared with an α-substituted compound is due in these cases, apparently, to the fact that, in spite of the lower effective charge on the β-substituent, the point distribution of charges of type (II) corresponds to a greater moment [formula (III.23)] than for point distribution (I). Examples of concrete calculations agreeing with this conclusion can be found in Forster and Nishimoto's paper [78].

For disubstituted naphthalenes, the following relation is satisfied fairly well [80]

$$\mu_{1.4} > \mu_{2.3} > \mu_{2.1} > \mu_{1.2}$$

where the index shows the position of the substituents. The question of the conjugation of substituents in various positions of the naphthalene ring is particularly interesting. Richards and Walker [81] have made a detailed study of the dipole moments of a series of isomeric nitronaphthalenes and have calculated the interaction moments of the amino and nitro groups in each isomer by a method analogous to that described for p-disubstituted benzenes. The results that they obtained are given in Table 44.

The values of the interaction moments show that the greatest conjugation between the nitro and amino groups in the naphthalene nucleus is achieved when these groups are present in the 1,4 positions. In this case, the interaction moment is even higher than for the corresponding p-derivative of benzene (see Table 39). Appreciable conjugation between the nitro and amino groups also takes place when they are located in the 1,7, 2,6, 2,7, and 2,8 positions. These data are in good agreement with ideas on the conjugation of substituents in the naphthalene series based on a study of reactivity. The magnitudes of μ_{int} from Table 44 correlate with the σ_{ij} constants of the naphthalene series obtained by Dewar and Grisdale [48].

Derivatives of other polynuclear aromatic compounds have been studied considerably less than the compounds of the naphthalene series. For those few 1- and 2-monosubstituted anthracenes the dipole moments of which have been determined, it is regularly found, as for the naphthalenes, that $\mu_2 > \mu_1$. Voluminous substituents in position 9 are repelled from the direction of the 1,8-hydrogen atoms. Because of this, mesomeric interaction with the nuclei is weakened, which explains the comparatively low dipole moment of 9-nitroanthracene (3.69 D [82]), approximately equal to the moment of nitrodurene (see Table 37).

Aromatic Heterocyclic Compounds

The group moments of substituents in heterocyclic compounds differ substantially from their values in heterocyclic conjugated compounds. Thus, the latter may serve as characteristics of the capacity of the corresponding heteryl system for electronic and spatial interaction with the substituents. Pyridine derivatives have been studied in most detail in this respect.

Katritzky, Sutton, et al. [47, 70, 83] have calculated the mesomeric moments of the substituents R in the 4-position of pyridine (I), its N-oxide (II), and its complex with boron trichloride:

$$\mu_M = \mu_R - \mu_H - \mu_{AlkR} \qquad (V.26)$$

The results of the determinations and calculations are given in Table 45.

A comparison of the mesomeric moments of substituents in compounds (I)-(III) with the mesomeric moments of the substitu-

TABLE 45. Mesomeric Moments of Substituents in Pyridine
and Its Derivatives [47, 70, 83]

Substituent R	I		II		III	
	μ_{exp}, D	μ_M, D	μ_{exp}, D	μ_M, D	μ_{exp}, D	μ_M, D
$N(CH_3)_2$	4.31	2.27	6.76	2.74	—	—
NH_2	3.96	1.55	—	—	—	—
OCH_3	2.96	1.16	5.08	1.39	8.86	1.35
Cl	0.78	0.57	2.82	0.59	6.71	1.02
CH_3	2.61	0.39	4.74	0.50	8.37	0.50
H	2.22	—	4.24	—	7.70	—
$COOC_2H_5$	2.53	−0.47	3.80	−0.93	7.74	−0.17
$COCH_3$	2.41	−0.21	3.19	−0.62	—	—
CN	1.65	−0.27	1.22	−1.50	4.20	−0.10
NO_2	1.58	−0.55	0	−0.99	—	—

ents in the benzene series leads to interesting conclusions. The
mesomeric moments of 4-substituents of the +M type in pyridine
are higher than in the benzene series. This is explained by the
polarization of the pyridine ring itself by the electronegative heteroatom and the induction in position 4 of some additional charge
under the influence of which the donor capacity of the +M substituent rises. The magnitude of the additional charge in the 4-position
of pyridine N-oxide and pyridine borotrichloride is obviously even
higher than in pyridine itself. This corresponds to the increase in
the mesomeric moments of the 4-substituents in the sequence (I)-
(II)-(III).

Conversely, for 4-substituents of the −M type, the mesomeric
moments in compounds (I) and (III) are lower than for substituents
in the benzene series, which agrees with the explanation given
above. Derivatives of type (II), in which the mesomeric moments
of −M substituents are greater (in absolute magnitude) than for
substituents in the benzene series, form an exception. The cause
of this is the capacity of the electronic system of pyridine N-oxide
for undergoing electromeric polarization of the type (IV)

which can also be described as a contribution of the limiting structure (V) [47]. This capacity is obviously lacking in compounds (I)
and (III).

The capacity of a substituent for conjugation with the heteryl system depends on the position of the substituent. For compounds of the pyridine series, this was shown by Cumper et al. [84-86] in several series of derivatives. It was established, for example, that the conjugation with the pyridine nucleus of substituents in position 3 was considerably lower than in positions 2 and 4.

The question of the interaction of ± M substituents with conjugated heteryl systems and the influence of this interaction on polar properties has so far been studied extremely unsystematically. To a certain extent, this is explained by the difficulties of synthesizing suitable reaction series of compounds. Reference can be made only to isolated papers of the dipole moments of derivatives of quinoline [85], indole [87], furan [71, 88], and thiophene [89].

Some other aspects of the application of the dipole moment to the study of the structure and properties of heterocyclic compounds are considered in a short review by Walker [90].

Nonaromatic Unsaturated Systems

Values of the dipole moments of a number of derivatives of unsaturated systems are given in Table 46. The moments of compounds of the ethylene series are lower than those of the corresponding substituted benzenes. Since the moments of the C −H bonds in ethylene and benzene are approximately equal (see Chapter III), this may be connected with the greater mesomeric interaction of +M substituents in the ethylene series than in the benzene series. Conversely, −M substituents possess mesomeric moments that are larger in absolute magnitude in the benzene series.

Similar comparisons lose their force on passing to a consideration of acetylenic derivatives. The high electronegativity of sp-hybridized carbon should lead to considerable changes in the magnitudes and even in the directions of the bond moments. For example, the zero dipole moment of bromoacetylene shows a reversal of the polarity of the C −Br bond in this compound as compared with aliphatic and olefinic derivatives: here the positive pole of the dipole is the bromine atom.

Using methods of correlation analysis, Charton [91] has shown that a considerable contribution to the group moment of substituents in unsaturated systems is made by the effect of conjuga-

TABLE 46. Dipole Moments of Derivatives of Saturated
Compounds of the Type R−X (in D) [20, 91]

R	X				
	−CH=CH₂	−CH=CH−CH=CH₂	−CH=C=CH₂	−C≡CH	−C≡C−CH=CH₂
H	0	0	0.20	0	0.80
F	1.43	—	—	0.75	—
Cl	1.44	—	—	0.44	—
Br	1.41	—	1.50	0	—
I	1.26	—	—	—	—
CH₃	0.36	0.68	0.40	0.75	0.93 [93]
CH₂Cl	1.08	—	2.02	1.65	—
CF₃	2·45	—	—	2.36	—
C₆H₅	0.13	—	—	0.73	0.27
OC₂H₅	1.27	—	—	1.98	—
SC₂H₅	1.36	—	—	—	1.51 [92]
CN	3.89	3.90	—	3.6	—
NO₂	3.41	—	—	—	—

tion with the unsaturated grouping. This effect is characteristic
for all the unsaturated systems given in Table 46 and is largest
for derivatives of the acetylene series. Petrov et al. have also
come to the conclusion that conjugation exists between substituents
adjacent to a triple bond in 1,3-enynes [92].

The question of the electronic conductivity of unsaturated
groupings is interesting. The correlation treatment [91] of the dipole moments of 1,2-disubstituted ethylenes and acetylenes in reaction series of the type R−M−Y (where Y always represents the
same substituent) leads to the conclusion that the conductivities of
the trans-vinylene, vinylidene, and acetylene systems are approximately the same and close to the conductivity of the p-phenylene
grouping. This conclusion was made, however, on the basis of a
comparatively small number of data and is not in agreement with
information on the reactivity of compounds of this type [1].

3. Dipole Moments and Intramolecular

Polarization

This question has already been touched on in our discussion
of the interaction moments of substituents in the benzene and naphthalene series, but it is not exhausted by the examples given. Below we shall consider the most characteristic structural features
responsible for a high polarity of the molecules of organic compounds.

TABLE 47. Dipole Moments of
Nonalternant Hydrocarbons

Hydrocarbon	μ_{exp} , D	μ_{calc} , D	
		HMO	SCF MO
Methylenecyclo-propene	—	5,0 [94]	1.21 [97]
Fulvene	1.1 [95, 98]	4.7 [96]	1.13 [101]
Azulene	1.0 [98]	6.9 [96]	1.7 [100]
Acenaphthylene	0.93 [99]	2.8 [104]	—
Acepleiadylene	0.49 [102]	5.7 [104]	—
Heptafulvene	0.5 [103]	2.35 [104]	—

Nonalternant Hydrocarbons and Their Derivatives

It has already been stated that nonalternant conjugated cyclic hydrocarbons possess considerable dipole moments. The reason for this is the peculiar symmetry of their wave functions, leading to the situation, unlike that of the alternant hydrocarbons, that the π-electronic charges on the atoms and, consequently, the dipole moments of the molecules are not equal to zero. This result, predicted by means of the simple LCAO MO method, is confirmed by the experimental data (Table 47). In addition to the experimental magnitudes, the table also gives values calculated theoretically by Hückel's simple MO method (HMO) and more rigorously taking self consistency into account (SCF MO).

The existence of dipole moments in nonalternant hydrocarbons may be substantiated in a somewhat simplified but qualitatively correct manner as a consequence of the distribution of the charges caused by the tendency of each conjugated cyclic system to assume the aromatic electronic configuration with $4n + 2$ π-electrons. In agreement with this, three-membered and seven-membered rings will eject the "superfluous" electron into the periphery of the molecule and five-membered rings, conversely, will strive to acquire the electron necessary for a sextet. Thus, it may be expected that the molecules of nonalternant hydrocarbons will be polarized in such a way that the positive pole of the dipole is localized in the 3- and 7-membered rings and the negative pole in the 5-membered ring. This corresponds to a contribution of the following electronic structures to a description of the ground electronic states of methylenecyclopropene (I), fulvene (II), azulene (III), and heptafulvene (IV):

This conclusion is confirmed by determinations of the dipole moments of a series of derivatives of nonalternant systems. For example, for 6,6-diphenylfulvene (V) a dipole moment of 1.34 D has been found. The moment of its dichloro derivative (VI) can be calculated on the assumption that the direction of the moment of the fulvene fragment is the same as, or is opposite to, that in compound (II).

In the first case, a value of about 0.3 D is obtained, since the vector of the dipole moment of fulvene is deducted from the sum of the vectors of the moments of the p-chlorophenyl residues. In the second case, the calculated value of the dipole moment of (VI) is 2.9 D. The experimental value [98] is obviously in favor of the polarization of fulvene (II) with the positive pole on the exocyclic carbon. A similar conclusion can be drawn on the basis of an analysis of the dipole moments of other substituted fulvenes [105].

In the case of azulene, the dipole moments of the 2-substituted derivatives (VII) also correspond to the assumption that the negative pole of the dipole is localized in the five-membered ring [106].

$X = Cl$, $\mu = 2.69\ D$
$X = CN$, $\mu = 5.68\ D$

In a consideration of the values of the dipole moments given, it must be borne in mind that the mesomeric interaction of the substituent in position 2 with the azulene system is somewhat higher than in the benzene series [107].

The intramolecular transfer of charge with the nonalternant system acting as a donor or acceptor of it becomes most considerable in those molecules in which such a system is conjugated with a heteroatomic grouping. The latter, when combined with 3- and 7-membered rings, must possess the properties of an electron acceptor, i.e., a high electron affinity, and when combined with a 5-membered ring, the properties of an electron donor, i.e., a low ionization potential. Molecules with such a structure have dipole moments considerably exceeding the additive values. Let us consider some examples of such polarized systems. Vol'pin, Koreshkov, and Kursanov [108] found a dipole moment of 5.08 D for 1,2-diphenylcyclopropenone (VIII), and Breslow et al., having confirmed this value in a recent paper, also determined the dipole moments of dipropylcyclopropenone (4.78 D) and cycloheptenocyclopropenone (4.66 D).

VIII   IX   X   XI

$\mu = 5.08\ D$ $\mu = 7.9\ D$

These values are 1.5-2 D higher than the additive values, which can be roughly estimated from the values of the moments of the cyclic ketones and cis-stilbene or the corresponding unsaturated hydrocarbons and a C_{sp^2} —H bond.

The existence of an additional moment is connected with the contribution of the polar structure (IX). An analogous polarization (XI) is observed in the case of another derivative of the pseudoaromatic methylenecyclopropene, the dinitrile (X) [110]. Here a comparison of the dipole moment of (X) with the dipole moment of diphenylmethylenemalononitrile (XII), in which the cyclopropene ring is replaced by an unsaturated system retaining the capacity only for ordinary π-conjugation, is particularly striking. The influence of the latter on the dipole moment is considerably less than the influence of intramolecular charge transfer (XI), although it is exerted in the same direction. This follows from comparisons of moments of (X), (XII), and (XIII) [110].

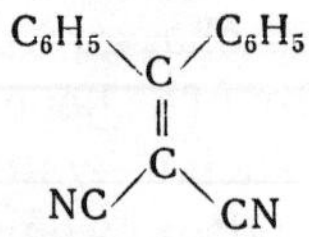

$$C_6H_5 \diagdown \underset{\|}{C} \diagup C_6H_5$$

XII

$\mu = 5.85\ D$

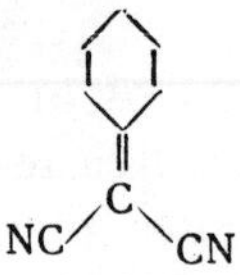

XIII

$\mu = 5.45\ D$

Identical electronic effects are characteristic for seven-membered cyclic homologs of compounds (VIII) and (X). This is shown by the dipole moments of tropone (XIV) [111] and of 8,8-heptafulvenedinitrile (XVI) [11, 113], which are polarized in the manners shown by formulas (XV) and (XVII), respectively.

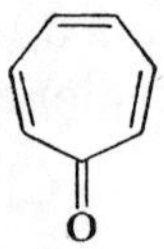

XIV

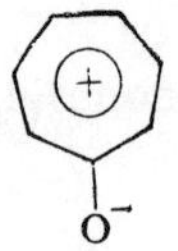

$\mu = 4.17\ D$

XV

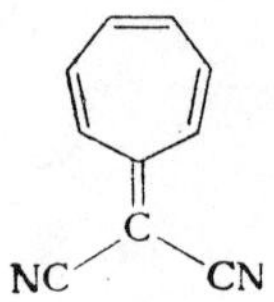

XVI

$\mu = 7.49\ D$

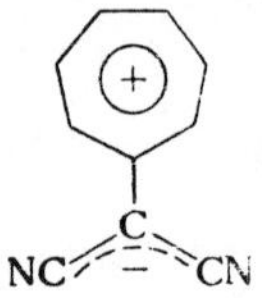

XVII

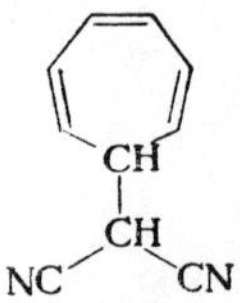

XVIII

$\mu = 4.12\ D$

A comparison of the dipole moments of (XVI) and its dihydro derivative (XVIII) [112] enables the contribution of intramolecular polarization of the type of (XVII) to the dipole moment of (XVI) to be evaluated. Dipole moments of the quinones of the sesquifulvalene series also considerably exceed the additive values. For sesquifulvalene-1,4-quinone (XIX) this difference is about 1 D and corresponds to the existence of intramolecular polarization of type (XX) [114]

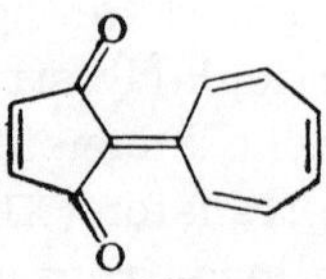

XIX

$\mu = 2.85\ D$

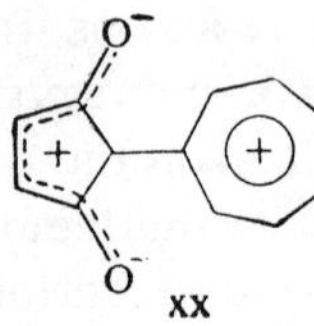

XX

A characteristic example of an intramolecularly polarized system containing a five-membered cyclopentadienyl ring is 6-dimethylaminofulvene (XXI).

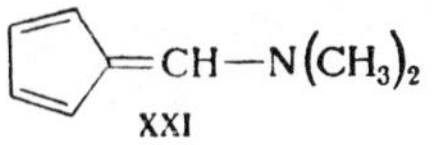
XXI

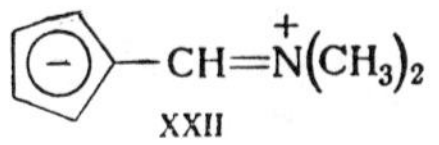
XXII

$\mu = 4.5\ D$

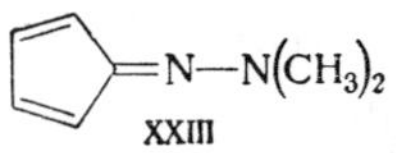
XXIII

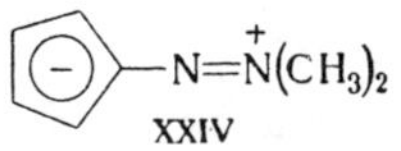
XXIV

$\mu = 3.3\ D$

The dipole moment found for (XXI) [115] is considerably greater than the additive value, which shows the role of the bipolar structure (XXII). The situation is analogous for the nitrogen analog of compound (XXI), the dimethylhydrazone of cyclopentadienone (XXIII). However, judging from the dipole moment [116], the contribution of the corresponding polarized structure (XXIV) is somewhat less in this case.

The clearly expressed capacity of the cyclopentadienyl system for completing its electron cloud to a sextet of electrons is shown in the formation of nitrogen and phosphorous ylides, the structure of which is determined by the transfer of an electron from nonbonding orbitals of the nitrogen and phosphorus to a lower bonding orbital of the cyclopentadienyl. As a result of this, the dipole moments of the cyclopentadienylides (XXV)-(XXVIII) are extremely high.

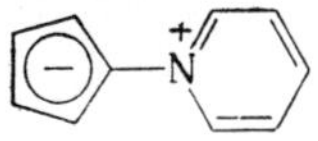
XXV

$\mu = 13.5\ D$

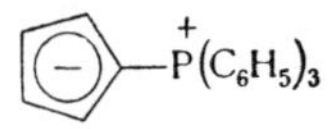
XXVI

$\mu = 7.0\ D$ [118]

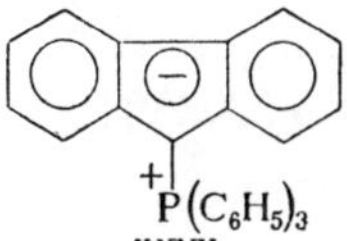
XXVII

$\mu = 7.09\ D$ [119]

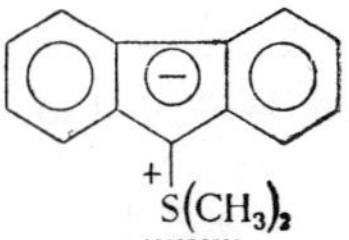
XXVIII

$\mu = 6.2 D$ [120]

The cycloheptatrienyl system or an analog of it may figure as the donor component attached to the cyclopentadienyl residue. This association, as follows from the discussion above, must lead to the intramolecular charge distribution that is observed, for example, in the case of azulene (III). For the heterocyclic isoelectronic analogs of azulene N-methyl-2-pyrindine (XXIX) and N-methyl-1-pyrindine (XXX), the SCF MO theory predicts a more pronounced intramolecular charge transfer than for azulene itself [121]

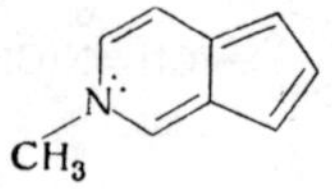

XXIX

$\mu_{calc} = 5.6\,D$

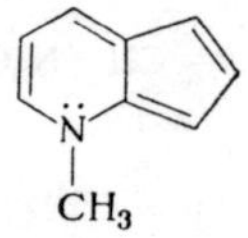

XXX

$\mu_{calc} = 4.9\,D$

In these compounds, as in azulene, the association of unsaturated rings with five and seven π-electrons takes place in the form of a condensed system. Another possibility, connection through a formal double bond, is realized in sesquifulvalene (XXXI) and its analogs.

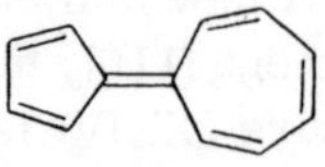

XXXI

XXXII

$\mu = 0.83\,D$

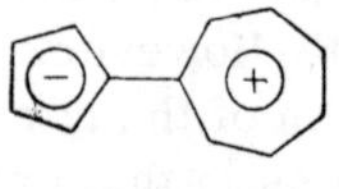

XXXIII

Sesquifulvalene itself has not so far been synthesized, but for its tetrabenzo derivative (XXXII) a small dipole moment has been found [122] which corresponds to the transfer of electrons from the seven-membered ring to the five-membered ring (XXXIII). The degree of intramolecular polarization of this type increases markedly when the cycloheptatrienyl residue in the molecule is replaced by a 1,2- or 1,4-dihydropyridyl residue isoelectronic with it [compounds (XXXIV) and (XXXV)]:

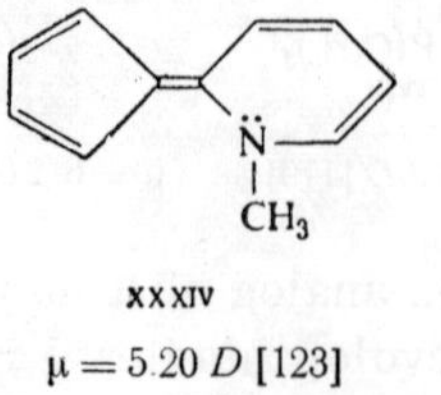

XXXIV

$\mu = 5.20\,D$ [123]

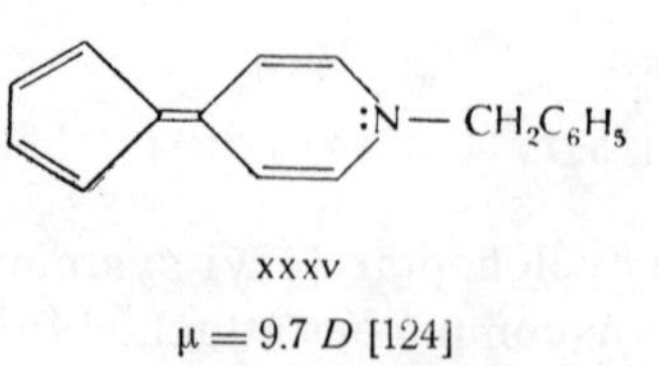

XXXV

$\mu = 9.7\,D$ [124]

Carbonyl, thiocarbonyl, and selenocarbonyl groups may serve as acceptors of the electrons of the dihydropyridine ring [125]. Compounds (XXXVI) of this type are analogous to tropone.

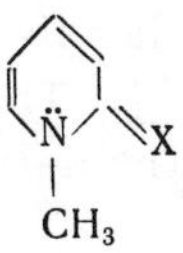

X	μ, D
O	4.04
S	5.26
Se	5.73

XXXVI

Attention is attracted by the progressive increase in the dipole moments of compounds (XXXVI) in the sequence X = O, S, Se. The same sequence is observed for other heterocyclic electronic analogs of tropone: the 5- and 3-antipyrines (XXXVII) and (XXXVIII) [126] and the γ-pyrones (XXXIX) and (XL) [127].

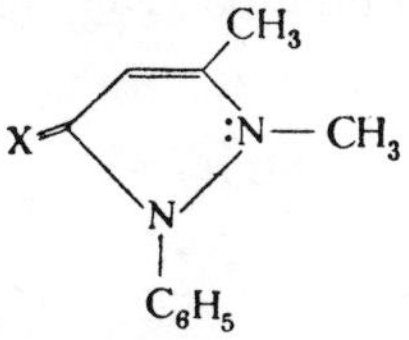

X	μ, D
NH	4.37
O	5.47
S	7.60
Se	7.91

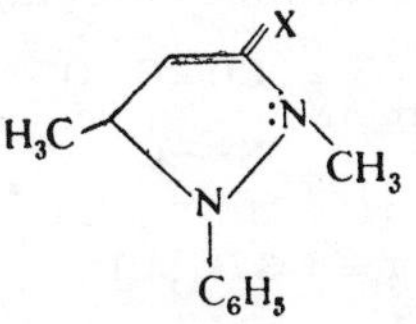

XXXVII

X	μ, D
NH	3.92
O	5.23
Se	8.17

XXXVIII

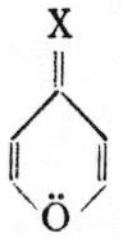

X	μ, D
O	3.72
S	4.08

XXXIX

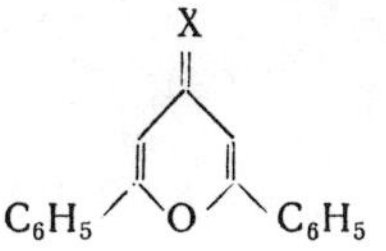

X	μ, D
O	4.74
S	5.31

XL

The reason is apparently the higher acceptor capacity of sulfur and selenium because of the presence in them of vacant d-orbitals the filling of which makes a wider distribution of charges possible in spite of the lower electronegativity of the elements of the higher periods. We may note that the increase in dipole moments when a carbonyl group is replaced by a thiocarbonyl group is also observed in the aromatic series for diaryl ketones [128], urea derivatives [129], merocyanines [130], etc.

Conversely, in the aliphatic series, where the effect of d,π - conjugation is not shown, the moments of thiocarbonyl compounds are considerably lower than those of the corresponding carbonyl compounds [131]. Thus, the dipole moment of cyclohexanone is 3.04 D, while the moment of cyclohexanethione is 1.70 D.

Mesoionic Compounds

A number of compounds exist the structure of which cannot be represented by any covalent scheme and can be illustrated only as a resonance hybrid of several bipolar forms. Such compounds are usually called mesoionic [132, 133]. The dipole moments of mesoionic compounds are usually extremely high, because of intramolecular polarization.

The best known example is formed by the sydnones. The investigation of the dipole moments of derivatives of N-phenylsydnone (I) with regular substituents in the aryl nucleus and on a cyclic carbon atom have shown, in agreement with structure (I) and other possible limiting structure, that the positive pole of the molecular dipole is located in the heterocycle [134-136]. This conclusion is based on the results of calculations by the MO method [133, 137].

$$C_6H_5-\overset{+}{N}\begin{matrix} \diagup CH=C-O^- \\ | \\ \diagdown N-O \end{matrix}$$
I
$\mu = 6.53\,D\,[135]$

$$C_6H_5-\overset{+}{N}\begin{matrix} \diagup CH-N-C_6H_5 \\ | \\ \diagdown N=C-\overset{-}{N}C_6H_5 \end{matrix}$$
II
$\mu = 7.2\,D\,[138]$

$$C_6H_5-\overset{+}{N}\begin{matrix} \diagup N=C-S^- \\ | \\ \diagdown C-S \end{matrix}$$
C_6H_5
III
$\mu = 8.8\,D\,[139]$

Other representatives of the mesoionic compounds are the derivative of the 1,2,4-triazole series (II), which is known under the name of nitrone, and the sulfur analog of sydnone (III). Of the polar resonance structures possible for them, structures (II) and (III) with a negative charge on the exocyclic phenylimine group and the sulfur atom have the most weight. In such structures, as in compound (I), the five-membered heterocycle acquires a stable six-π-electronic configuration, which is responsible for their stabilization and their high dipole moments.

The mesoionic compounds also include the azides, the structure of which corresponds to the resonance hybrid (IV), diazo compounds of type (V), and quinone diazides (VI):

$$C_2H_5-\overset{+}{N}=\overset{-}{N}=\overset{-}{N} \leftrightarrow C_2H_5-\overset{-}{N}-\overset{+}{N}\equiv N$$
IV
$\mu = 2.14\,D\,[140]$

$$\begin{matrix} C_6H_5 \\ \diagup \\ C_6H_5 \end{matrix}C=\overset{+}{N}=\overset{-}{N} \leftrightarrow \begin{matrix} C_6H_5 \\ \diagup \\ C_6H_5 \end{matrix}\overset{-}{C}-N\equiv\overset{+}{N}$$
V
$\mu = 1.42\,D\,[142]$

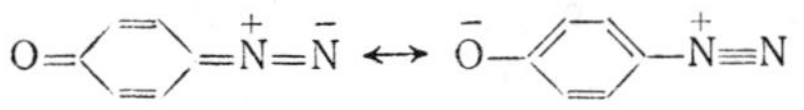

$$O=\left\langle\right\rangle=\overset{+}{N}=\overset{-}{N} \longleftrightarrow \overset{-}{O}-\left\langle\right\rangle-\overset{+}{N}\equiv N$$

VI

$\mu = 5.0\ D$ [141]

Judging from the dipole moments, it may be considered that in the case of the azides and diazo compounds the contributions of the structures of opposite polarization do not differ so strongly from one another as in the case of p-quinone diazide (VI).

Other typical mesoionic compounds are the betaines. Only a few of them possess a solubility in nonpolar or sparingly polar solvents that is sufficient for a determination of their dipole moments. In view of this, attempts have been made to determine the dipole moments of some betaines (for example, internal salts of amino acids) from the dependence of the dielectric constant of the aqueous solutions on the concentration [144].

Nevertheless, for some betaines, it has been possible to obtain values of their dipole moments for measurements carried out in benzene. We may refer to the azomethine imine (VII), the pyridinium-N-phenol betaine (VIII), and the betaine (IX).

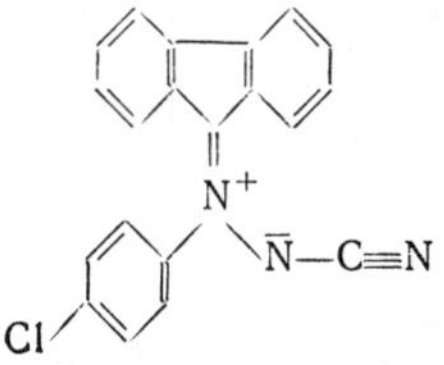

VII

$\mu = 6.7\ D$ [142]

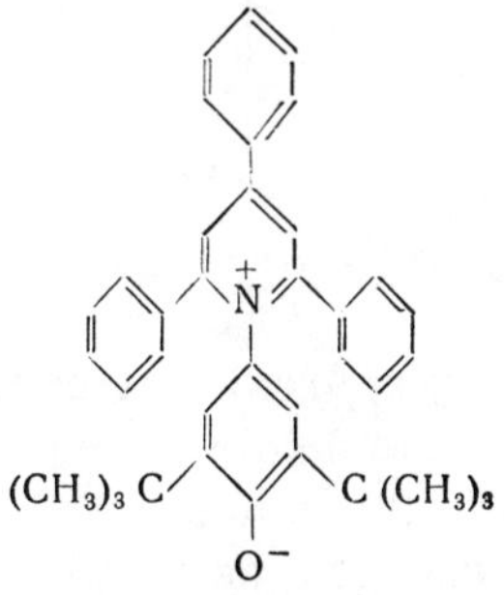

VIII

$\mu = 14\ D$ [143]

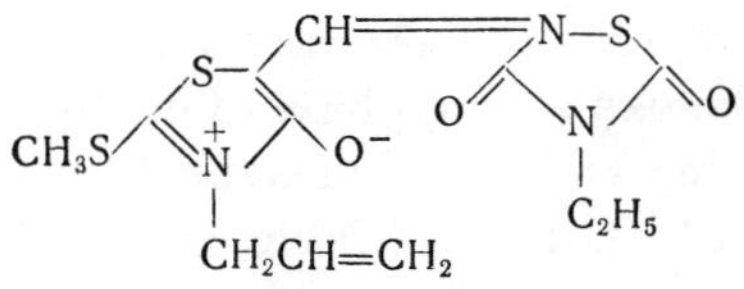

IX

$\mu = 7.11\ D$ [175]

TABLE 48. Dipole Moments of the Vinylogs $D-(CH=CH)_n-A$
(Benzene, 25°C)

D	A	n	μ, D	Literature
$N(CH_3)_2$	CHO	0 1 2 3 4	3.86 6.24 7.67 8.24 8.50	[176]
C_6H_5	CHO	0 1 2 5	2.96 3.59 3.80 4.16	[145] [145] [146] [146]
	$=NC_6H_5$	0 * 1 2	2.37 4.17 5.32	[147]
		0 1 2	3.46 6.28 7.61	[148]
C_6H_5NH		0 1 2	4.27 5.85 6.28	[148]

* Repeating link $=(CH-CH)_n=$

From an approximate experimental determination, the π-component of the dipole moment of (VIII) is 18–20 D [143]. This is fairly close to the value of 27 D calculated for the complete separation of the charges, as in structure (VIII).

Compounds with an Extended Chain of Conjugated Bonds

Extremely high dipole moments are found for compounds the molecules of which possess a long chain of conjugated bonds separating electron-donating and electron-accepting groups, for example of the type $D-(CH=CH)_n-A$ (Table 48).

The main reason for the rise in polarity with an increase in the length of the chain of conjugation in this case is apparently the increase in the arm of the dipole separating the positive and negative charges localized, respectively, on the donor and acceptor groups. As can be seen from molecular diagrams of dimethylformamide and its vinylogs calculated by the LCAO MO method,[*] an elongation of the chain of conjugation is accompanied by some decrease in the effective charges on the nitrogen and oxygen atoms, but the dipole moments calculated from the electronic configurations given rise. At the same time, the dipole moments tend to an upper limit (for $Me_2N-(CH=CH)_nCHO$ $\mu_{lim} = 12.8$ D), which corresponds to the experimentally observed value (see Table 48).

I

$$\underset{Me_2N}{\overset{+0.167}{}}\underset{}{\overset{+0.409}{-\!-\!-C}}\overset{O^{-0.576}}{\underset{H}{\diagup\!\!\diagdown}}\qquad \mu_\pi = 4.3\ D$$

II

$$\underset{Me_2N}{\overset{0.143}{}}\overset{+0.220}{-\!-\!-CH}\underset{}{\overset{-0.131}{\diagup}}\overset{+0.329}{CH-\!-\!-C}\overset{O^{-0.561}}{\underset{H}{\diagup\!\!\diagdown}}\qquad \mu_\pi = 7.3\ D$$

III

$$\underset{Me_2N}{\overset{+0.130}{}}\overset{+0.159}{-\!-\!-CH}\overset{-0.115}{\diagup}\overset{+0.148}{CH-\!-\!-CH}\overset{-0.085}{\diagup}\overset{+0.317}{CH-\!-\!-C}\overset{O^{-0.553}}{\underset{H}{\diagup\!\!\diagdown}}\qquad \mu_\pi = 9.6\ D$$

IV

$$\underset{Me_2N}{\overset{+0.123}{}}\overset{+0.129}{-\!-\!-CH}\overset{-0.114}{\diagup}\overset{+0.087}{CH-\!-\!-CH}\overset{-0.056}{\diagup}\overset{+0.135}{CH-\!-\!-CH}\overset{-0.068}{\diagup}\overset{+0.315}{CH-\!-\!-C}\overset{O^{-0.548}}{\underset{H}{\diagup\!\!\diagdown}}\qquad \mu_\pi = 11.2\ D$$

V

$$\underset{Me_2N}{\overset{+0.111}{}}\overset{+0.112}{-\!-\!-CH}\overset{-0.114}{\diagup}\overset{+0.058}{CH-\!-\!-CH}\overset{-0.053}{\diagup}\overset{+0.075}{CH-\!-\!-CH}\overset{-0.039}{\diagup}\overset{+0.131}{CH-\!-\!-CH}\overset{-0.060}{\diagup}\overset{+0.316}{CH-\!-\!-C}\overset{O^{-0.545}}{\underset{H}{\diagup\!\!\diagdown}}\qquad \mu_\pi = 12.3\ D$$

The degree of intramolecular polarization depends greatly on the nature of the donor and acceptor groupings. Thus, the merocyanine dye (VI) possesses a moment corresponding to almost complete charge transfer of type (VII), while in the case of the merocyanines given in Table 48 this transfer is inhibited to a considerable extent.

---

[*] Hückel's simple method was used with Streitwieser's parameters [96] and with the alternation of the bonds ($k_{C-C} = 0.9$; $k_{C=C} = 1.1$) taken into account. The lengths of the bonds were taken from published tables [17]. The dipole moments were calculated for the s-trans configurations (I)-(V). The contribution of other configurations was not taken into account.

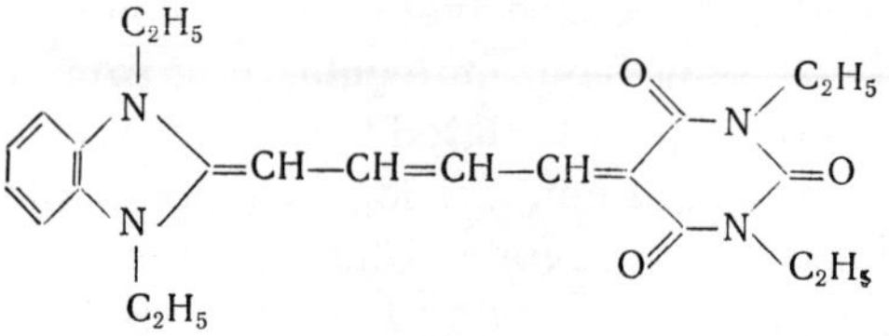

VI

$\mu = 17.7\ D$ [147]

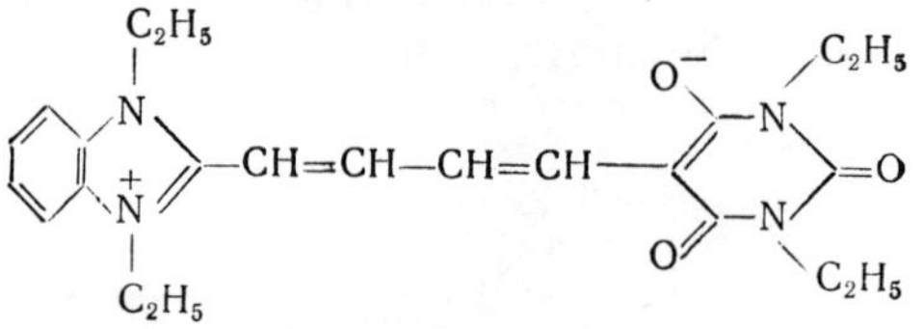

VII

The theory of solvatochromy [149] (the change in the color of dyes on passing from nonpolar solvents to polar solvents and conversely) is connected with ideas on the polarization of cyanine dyes of this type. For the majority of dyes, a deepening of the coloration with an increase in the polarity of the solvent is characteristic, this being connected with the stabilization of the intramolecularly polarized form. So far, however, the dipole moments of the cyanines have been determined only in nonpolar solvents. It would be interesting to supplement the results obtained by measurements in polar media.

Molecules with Coordination Bonds

A high polarity of an organic compound may be due to the presence in its molecule of a coordination (semipolar) bond. Thus, the dipole moment of trimethylamine N–oxide (I) is 4.9 D [120], and is determined almost completely by the moment of the coordination bond.

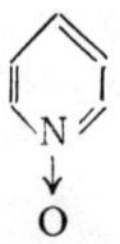

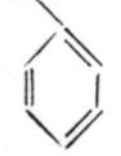

If the transfer of an electron from a nitrogen atom into the p–orbital of the oxygen in the coordination bond were total, so that the N–oxide had the bipolar structure (II), the dipole moment would be ~5.9 D [150]. A comparison of this figure with the experimen-

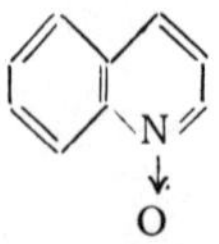

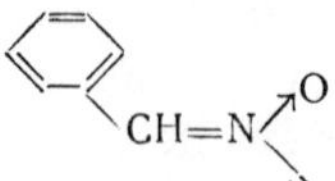

tal value shows that the effective charges on the nitrogen and oxygen are in fact somewhat less than is predicted by structure (II).

Even lower as compared with those calculated for the bipolar
forms are the dipole moments of those N-oxides the molecules of
which contain double bonds and aromatic rings. For example, the
dipole moment of pyridine N-oxide (III) is 4.18 D [151], that of
quinoline N-oxide (IV) 4.00 D [152], and that of the nitrone (V)
3.37 D [150]. The reason is the dispersal over the conjugated system of charges that are localized on the coordination bond in the
saturated structure (I).

N-Oxides are the most widely studied but not, of course, by
any means the only examples of compounds with coordination bonds
possessing high dipole moments. We may also mention a few compounds of other similar types: the sulfoxides (VI) and (VII) and
the selenoxide (VIII), the arsine oxides (IX) and the phosphine oxides, the stilbine sulfides (X), etc.

$$(C_2H_5)_2 S \longrightarrow 0 \quad (C_6H_5)_2 S \longrightarrow 0 \quad (C_6H_5)_2 Se \longrightarrow 0$$
$$\text{VI} \qquad\qquad \text{VII} \qquad\qquad \text{VIII}$$
$$\mu = 3.85\,D\,[153] \quad \mu = 4.00\,D\,[154] \quad \mu = 4.44\,D\,[155]$$

$$(C_6H_5)_3 As \longrightarrow 0 \qquad (C_6H_5)_3 Sb \longrightarrow S$$
$$\text{IX} \qquad\qquad\qquad \text{X}$$
$$\mu = 5.50\,D\,[155] \qquad\qquad \mu = 5.40\,D$$

By making use of a vectorial additive scheme, from the dipole moments of the compounds given above and others Smyth [156]
has calculated the moments of several fundamental coordination
bonds found in the molecules of organic compounds. The results
of his calculations, based mainly on the data of Sutton et al. [120]
and Jensen [155] and also on the values of the moments of coordination bonds obtained in later work, are given in Table 49.

It must, however, be borne in mind that the values of the moments of the coordination bonds given in Table 49 are an extremely
rough measure of their polarity and depend on the structure of the
compound from the dipole moment of which they were evaluated.

As a rule, the highest polarity is possessed by coordination
bonds in complex organic ligands with metals. The presence of
such bonds explains the high dipole moments of compounds (XI)-
(XVI).

TABLE 49. Dipole Moments of Coordination Bonds

Bond	Bond moment, D	Bond	Bond moment, D
N → O	4.3	Se → O	3.1
P → O	2.9 [157]	Te → O	2.3
P → S	3.1	N → B	2.55 [159]
P → Se	3.2	P → B	4.4
As → O	4.2	O → B	3.6
Sb → S	4.5	S → B	3.8
S → O	3.0 [158]		

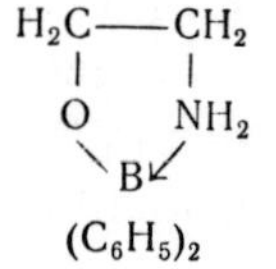

$$H_2C \text{---} CH_2$$

XI

$\mu = 8.7\ D$ [160]

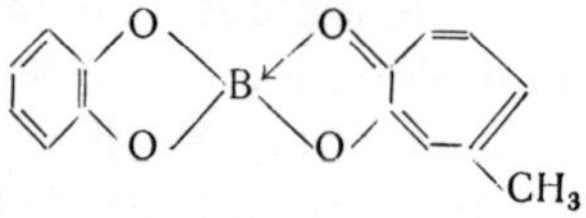

XII

$\mu = 8.24\ D$ [161]

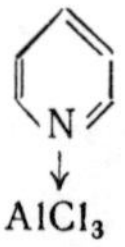

XIII

$\mu = 8.02\ D$ [162]

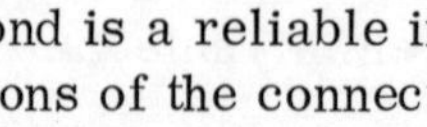

XIV

$\mu = 4.89\ D$ [163]

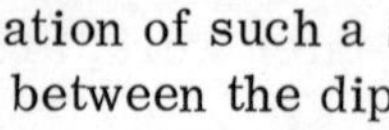

XV

$\mu = 6.49\ D$ [164]

$(C_6H_5)_3\ P \longrightarrow BCl_3$

XVI

$\mu = 7.011\ D$ [120]

Frequently, the observation of a high dipole moment in a compound one of the possible structures of which has a coordination bond is a reliable indication of such a structure [161, 165]. Questions of the connection between the dipole moments and structure of coordination compounds with organic ligands have been discussed in reviews [166, 167].

Transannular Interaction

Donor–acceptor interaction with the appearance of a polar coordination bond in a molecule is also possible between suitable atoms separated from one another by several bonds but, because of features of the molecular geometry, spatially close enough for

the existence of such an interaction. This type of intramolecular
interaction, which is called transannular [168, 169], is exhibited in
the appearance of a fairly high additional moment directed along
the line of the transannular coordination bond.

One of the molecular systems the conformation of which is
capable of forming transannular coordination bonds is the 1-azacy-
clooctan-5-one system, e.g., in the cyclic amino ketone (I). The
dipole moment calculated for the most stable conformation is
1.3 D lower than the experimental value. The assignment of the
additional moment to the dipole of the transannular bond corresponding to a contribution of the bipolar structure (II) is confirmed
by IR-spectroscopic data and the features of the reactivity of
compound (I) [170].

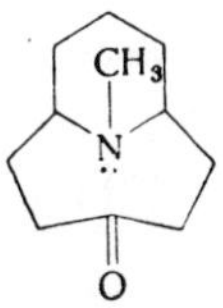

$\mu = 4.87 D$

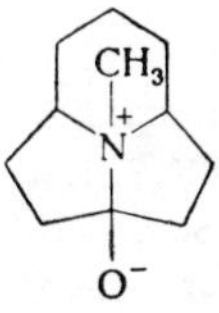

Similar results have been obtained for 1-oxa- and 1-thiacy-
clooctan-5-ones [171].

The dipole moment calculated for the bipolar structure (II)
is about 11 D. Thus, the transannular interaction of the amino and
keto groups in structure (I) is accompanied by only slight charge
transfer. The transannular transfer of charges is considerably
more complete in the eight-membered rings of the atranes, heteroorganic compounds with coordination bonds of the N$\rightarrow$Z type,
where Z = B, Si, etc.

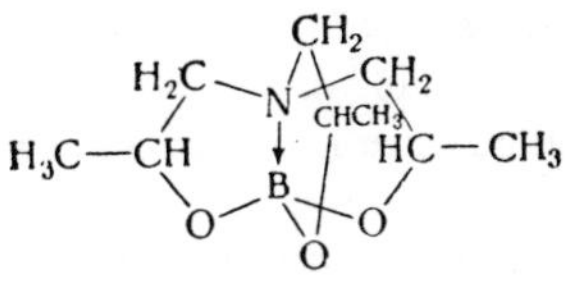

III
$\mu = 6.69 D$ [172]

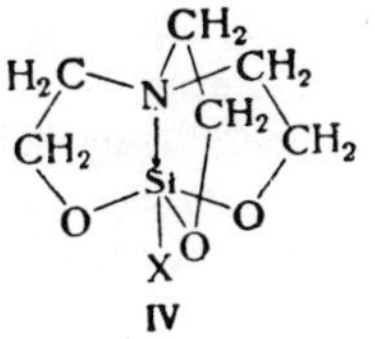

IV
X = CH$_3$, $\mu = 5.30 D$ [173]
X = OC$_6$H$_5$, $\mu = 7.13 D$ [173]

In the silatranes (IV), the calculated moments of the N$\rightarrow$Si
coordination bond are about 5.2 D [173]. A comparison of this

value with the moment of (IV) calculated for the complete transfer of an electron from the nitrogen to the silicon and equal to 8.6 D for a $\overset{+}{N}-\overset{-}{Si}$ distance of 1.8 Å shows the extremely polar nature of this transannular bond. It is interesting that the high moments of the silatranes (IV) show unambiguously the correctness of their spatial "concave-convex" structure in which the nitrogen and silicon atoms are at a distance permitting the formation of a coordination bond between them. From spatial considerations, such a bond is impossible in the alternative unstrained "biconvex" structure. The dipole moments calculated for the latter do not exceed 1 D, which excludes the possibility of their existence [173, 174].

Literature Cited

1. Yu. A. Zhdanov and V. I. Minkin, Correlation Analysis in Organic Chemistry, Izd. Rostov-on-Don State University (1966).
2. H. M. Smallwood and K. F. Herzfeld, J. Am. Chem. Soc., 52:1919 (1930).
3. F. Frank, Proc. Roy. Soc., A152:171 (1936).
4. J. W. Smith, Electric Dipole Moments, Butterworths, London (1955), Chap. 7.
5. B. A. Arbuzov and A. N. Vereshchagin, Izv. Akad. Nauk SSSR, Ser. Khim., 1004 (1964).
6. A. N. Vereshchagin and S. G. Vul'fson, Teoret. Éksperim. Khim., Akad. Nauk UkrSSR, 1:305 (1965).
7. L. G. Groves and S. Sugden, J. Chem. Soc., 1937:1992.
8. A. N. Vereshchagin, Author's Abstract of Candidate's Thesis, Kazan' State University (1964).
9. C. P. Smyth and K. B. M. McAlpine, J. Chem. Phys., 1:190 (1933).
10. A. E. Remick, J. Chem. Phys., 9:653 (1941).
11. R. P. Smith, T. Ree, J. L. Magee, and H. Eyring, J. Am. Chem. Soc., 73:2263 (1951).
12. R. P. Smith and E. M. Mortensen, J. Am. Chem. Soc., 78:3932 (1956).
13. W. Kauzmann, Quantum Chemistry, Academic Press, New York (1957).
14. K. G. Denbigh, Trans. Faraday Soc., 36:936 (1940).
15. C. W. Bunn and R. P. Daubeny, Trans. Faraday Soc., 50:1173 (1954).
16. R. J. W. LeFevre, Advan. Phys. Chem., Vol. 3 (1965).
17. Tables of Interatomic Distances and Configuration in Molecules and Ions, Spec. Publ., No. 11, Chem. Soc., London (1958).
18. K. Antos, A. Martvoni, and P. Kristian, Collection Czech. Chem. Commun., 31:3737 (1966).
19. V. N. Krishnamurthy and S. Soundararajan, J. Inorg. Nucl. Chem., 27:2341 (1965).
20. O. A. Osipov and V. I. Minkin, Handbook on Dipole Moments, Izd. Vysshaya Shkola (1965).
21. W. H. Flygare, A. Narath, and W. D. Gwinn, J. Chem. Phys., 36:200 (1962).
22. N. D. Sokolov, Usp. Khim., 36:2195 (1967).

23. R.J.W. LeFevre and B.J. Orr, J. Chem. Soc., 1966B:37.

24. R.J.W. LeFevre and A.J. Williams, J. Chem. Soc., 1965:4185.

25. R.J.W. LeFevre and B.J. Orr, J. Chem. Soc., 1965:2499, 5349.

26. L. Pierce and J.F. Becker, J. Am. Chem. Soc., 88:5406 (1966).

27. M. Kesler, Croat. Chem. Acta, 35:101 (1963).

28. D.R. Lide and D.E. Mann, J. Chem. Phys., 29:914 (1958).

29. G. Rothschild, J. Chem. Phys., 45:1214 (1966).

30. S. Weiss, J. Phys. Chem., 70:3146 (1966).

31. K. Syamalamba and D. Premaswarup, Indian J. Pure Appl. Phys., 2:11 (1964).

32. A.E. Lutskii, V.T. Alekseeva, and V.P. Kondratenko, Zh. Fiz. Khim., 35:1706 (1961).

33. A.E. Littlejohn and J.W. Smith, J. Chem. Soc., 1953:2456; 1954:2552.

34. G.C. Hampson and A. Weissberger, J. Chem. Soc., 1936:393.

35. O. Bastiansen and O. Hassel, Acta Chem. Scand., 1:489 (1947).

36. L.E. Sutton, Proc. Roy. Soc., A133:668 (1931).

37. K.B. Everard and L.E. Sutton, J. Chem. Soc., 1951:2816-2818, 2821, 2826.

38. L.E. Sutton, Determination of Organic Structures by Physical Methods, New York (1955), p. 272.

39. C.K. Ingold, Structure and Mechanism in Organic Chemistry, G. Bell & Sons, London (1953).

40. O. Exner, Collection Czech. Chem. Commun., 25:735 (1960).

41. A.R. Katritzky and P. Simmons, J. Chem. Soc., 1959:2051.

42. H. Lumbroso, Bull. Soc. Chim. France, 1955:643.

43. M.G. Voronkov, Izv. Akad. Nauk SSSR, Ser. Khim., 1688 (1961).

44. A. Audsley and F.R. Goss, J. Chem. Soc., 1942:497.

45. H. Lumbroso and G. Dumas, Bull. Soc. Chim. France, 1955:651.

46. H. Kofod, L.E. Sutton, P.E. Verkade, and B.M. Wepster, Rec. Trav. Chim., 78: 790 (1959).

47. A.R. Katritzky, E.W. Randall, and L.E. Sutton, J. Chem. Soc., 1957:1769.

48. M.J.S. Dewar and P.J. Grisdale, J. Am. Chem. Soc., 84:3539, 3548 (1962).

49. J. Trotter, Canad. J. Chem., 37:1487 (1959).

50. J.W. Smith, J. Chem. Soc., 1961:81.

51. R. Nakashima, Bull. Chem. Soc. Japan, 34:1740 (1961).

52. M.J. Aroney, M.G. Corfield, and R.J.W. LeFevre, J. Chem. Soc., 1964:648, 2954.

53. E.A. Brause, F. Sondheimer, and W.F. Forbes, Nature, 173:117 (1954).

54. V. Baliah and M. Uma, Tetrahedron, 19:455 (1963).

55. N. Mataga, Bull. Chem. Soc. Japan, 36:1607 (1963).

56. S. Forsen and T. Alm, Acta Chem. Scand., 19:2027 (1965).

57. S. Nagakura, M. Kojima, and Y. Maruyama, J. Mol. Spectr., 13:174 (1964).

58. H. Labhart and G. Wagniere, Helv. Chim. Acta, 46:1314 (1963).

59. O. Exner, Proceedings of a Conference on "Correlation Equations in Organic Chemistry" [in Russian], No. 1, Tartu (1964), p. 67.

60. R.J.K. Marsden and L.E. Sutton, J. Chem. Soc., 1936:599, 1383.

61. A.E. Lutskii, L.M. Yagupol'skii, and E.M. Obukhova, Zh. Obshch. Khim., 34: 2641 (1964).

62. G. Cilento, Chem. Rev., 60:147 (1960).

63. Yu. I. Naumov and V. I. Minkin, Zh. Fiz. Khim., 40:2569 (1966).
64. H. H. Huang and K. M. Hui, J. Organometal. Chem., 2:288 (1964).
65. H. Schindlbauer and G. Hajek, Ber., 96:2601 (1963).
66. K. A. Jensen and N. H. Bang, Liebigs Ann. Chem., 548:110 (1941).
67. V. I. Minkin, E. A. Medyantseva, and A. M. Simonov, Dokl. Akad. Nauk SSSR, 149:1347 (1963).
68. V. I. Minkin, Zh. Fiz. Khim., 41:556 (1967).
69. V. I. Minkin, Yu. A. Zhdanov, E. A. Medjantzeva, and Yu. A. Ostroumov, Tetrahedron, 23:3651 (1967).
70. A. J. Boulton, G. M. Glover, M. H. Hutchinson, A. R. Katritzky, D. J. Short, and L. E. Sutton, J. Chem. Soc., 1966 B:822.
71. L. M. Nazarova, Zh. Obshch. Khim., 32:1423 (1962).
72. V. N. Vasil'eva, Zh. Obshch. Khim., 35:218 (1965).
73. H. C. Longuet-Higgins, J. Chem. Phys., 18:265 (1950).
74. N. Nakata, Ber., 64:2059 (1931).
75. A. Parts, Z. Phys. Chem., B10:264 (1931).
76. A. E. Lutskii and L. A. Kochergina, Zh. Fiz. Khim., 37:460 (1963).
77. J. Trotter, Acta Cryst., 13:95 (1960).
78. L. S. Forster and K. N. Nishimoto, J. Am. Chem. Soc., 87:1459 (1965).
79. M. Kojima and S. Nagakura, Bull. Chem. Soc. Japan, 39:1262 (1966).
80. A. E. Lutskii, L. A. Kochergina, and B. A. Zadorozhnyi, Zh. Fiz. Khim., 37:671 (1963).
81. J. H. Richards and S. Walker, Tetrahedron, 20:841 (1964).
82. K. Nakamura and R. Nakashima, Nippon Kagaku Zasshi, 83:226 (1962).
83. C. M. Bax, A. R. Katritzky, and L. E. Sutton, J. Chem. Soc., 1958:1254, 1769.
84. C. W. N. Cumper, R. F. A. Ginman, and A. I. Vogel, J. Chem. Soc., 1962:4518, 4525
85. C. W. N. Cumper, R. F. A. Ginman, D. G. Redford, and A. I. Vogel, J. Chem. Soc., 1963:1731.
86. D. J. W. Bullock, C. W. N. Cumper, and A. I. Vogel, J. Chem. Soc., 1965:5311.
87. H. Lumbroso and G. Pappalardo, Bull. Soc. Chim. France, 1961:1131.
88. L. M. Nazarova and Ya. K. Syrkin, Izv. Akad. Nauk SSSR, Ser. Khim., 35 (1939).
89. T. Shimozawa, Bull. Chem. Soc. Japan, 38:1064 (1965).
90. S. Walker, in: Physical Methods in the Chemistry of Heterocyclic Compounds, Ed. R. P. Katrizky, Academic Press, New York (1963).
91. M. Charton, J. Org. Chem., 30:552 (1965).
92. A. A. Petrov, S. I. Radchenko, K. S. Mingaleva, I. G. Savich, and V. B. Lebedev, Zh. Obshch. Khim., 34:1899 (1964).
93. A. A. Petrov, K. S. Mingaleva, and B. S. Kupin, Dokl. Akad. Nauk SSSR, 123:298 (1958).
94. G. Berthier and B. Pullman, Bull. Soc. Chim. France, 1949:457.
95. J. Thiec and J. Wiemahn, Bull. Soc. Chim. France, 1956:177.
96. A. Streitwiesser, Molecular Orbital Theory for Organic Chemists, Wiley, New York (1961).
97. A. Julg, J. Chim. Phys., 50:652 (1953).
98. G. W. Wheland and D. E. Mann, J. Chem. Phys., 17:264 (1949).

99. T. Ishiguro, T. Chiba, and N. Gotoh, Bull. Chem. Soc. Japan, 30:25 (1957).

100. A. Julg, Compt. Rend., 239:1498 (1954).

101. G. Berthier, J. Chem. Phys., 21:953 (1953).

102. D. Pitt, A. J. Petro, and C. P. Smyth, J. Am. Chem. Soc., 79:5633 (1957).

103. Ya. K. Syrkin and E. A. Shott-L'vova, Acta Physicochim. URSS, 19:379 (1944).

104. A. Pullman, B. Pullman, E. D. Bergmann, G. Berthier, E. Fischer, Y. Hirschberg, and J. Pontis, J. Chim. Phys., 48:359 (1951); 49:20 (1952).

105. G. Kresze and H. Goetz, Ber., 90:2161 (1957).

106. J. Kurita and M. Kubo, J. Am. Chem. Soc., 79:5460 (1957).

107. E. Heilbronner, Nonbenzenoid Aromatic Compounds [Russian translation], IL, Moscow (1953), Chap. 5.

108. M. E. Vol'pin, Yu. D. Koreshkov, and D. N. Kursanov, Izv. Akad. Nauk SSSR, Ser. Khim., 560 (1959).

109. R. Breslow et al., J. Am. Chem. Soc., 87:1326 (1965).

110. E. D. Bergmann and I. Agranat, J. Am. Chem. Soc., 86:3587 (1964).

111. Y. Kuruta, T. Nozoe, and M. Kubo, Bull. Chem. Soc. Japan, 26:242 (1953).

112. M. Yamakawa, H. Watanabe, T. Mukai, T. Nozoe, and M. Kubo, J. Am. Chem. Soc., 82:5665 (1960).

113. H. Weller-Feilchenfeld, I. Agranat, and E. D. Bergmann, Trans. Faraday Soc., 62:2084 (1966).

114. S. Katagiri, I. Murata, Y. Kitahara, and H. Azumi, Bull. Chem. Soc. Japan, 38:282 (1965).

115. K. Hafner, K. H. Vöpel, G. Ploss, and C. König, Ann. Chem., 661:52 (1963).

116. K. Hafner and K. Wagner, Angew. Chem., 75:1104 (1963).

117. D. N. Kursanov and N. K. Baranetskaya, Izv. Akad. Nauk SSSR, Ser. Khim., 341 (1958).

118. F. Ramirez and S. Levy, J. Am. Chem. Soc., 79:67 (1957).

119. A. W. Johnson, J. Org. Chem., 24:282 (1959).

120. G. M. Phillips, J. S. Hunter, and L. E. Sutton, J. Chem. Soc., 1945:146.

121. J. A. Berson, E. M. Evleth, and S. I. Manatt, J. Am. Chem. Soc., 87:2901 (1965).

122. B. Pullman et al., Bull. Soc. Chim. France, 1952:73.

123. W. D. Kumler, J. Org. Chem., 28:1731 (1963).

124. D. N. Kursanov, M. E. Vol'pon, and Z. N. Parnes, Khim. Nauka i Prom., 3:159 (1958).

125. M. H. Krackow, C. M. Lee, and H. G. Mautner, J. Am. Chem. Soc., 87:892 (1965).

126. A. D. Garnovskii, V. I. Minkin, I. I. Grandberg, and T. A. Ivanova, Khim. Geterotsikl. Soedin. (Collection, 1967).

127. M. Rolla, M. Sanesi, and G. Traverso, Ann. Chim. (Rome), 42:664 (1952); 44: 430 (1954).

128. A. Lüttringhaus and J. Grohmann, Z. Naturforsch., 10b:365 (1955).

129. H. G. Mautner and W. D. Kumler, J. Am. Chem. Soc., 78:97 (1956).

130. E. A. Shott-L'vova, Ya. K. Syrkin, I. I. Levkoev, and M. V. Deichmeister, Dokl. Akad. Nauk SSSR, 145:1321 (1962).

131. H. Lumbroso and C. Audrien, Bull. Soc. Chim. France, 1966:3201.

132. W. Baker and W. D. Ollis, Quart. Rev., 11:15 (1957).

133. C. Coulson, Valence, Oxford University Press, 2nd Ed. (1961).

134. R. A. W. Hill and L. E. Sutton, J. Chim. Phys., 46:244 (1949).

135. J. C. Earl, E. M. W. Leake, and R. J. W. LeFevre, J. Chem. Soc., 1947:2269.

136. I. Fischer, Nature (London), 165:239 (1950).

137. D. A. Bochvar and A. A. Bagatur'yants, Zh. Fiz. Khim., 39:1631 (1965).

138. F. L. Warren, J. Chem. Soc., 1938:1100.

139. K. A. Jensen and A. Friediger, Kgl. Danske Videnskab. Selskab, Mat-fys. Medd., 20:20 (1943).

140. E. A. Shott-L'vova and Ya. K. Syrkin, Dokl. Akad, Nauk SSSR, 87:639 (1952).

141. J. D. C. Anderson, R. J. W. LeFevre, and J. R. Wilson, J. Chem. Soc., 1949:2082.

142. N. V. Sidgwick, L. E. Sutton, and W. Thomas, J. Chem. Soc., 1933:406.

143. A. Schweig and C. Reichardt, Z. Naturforsch., 21a:1373 (1966).

144. W. Hückel, Theoretische Grundlagen der organischen Chemie, Akademische Verlag, Leipzig, 9th Ed. (1957).

145. E. Bramley and R. J. W. LeFevre, J. Chem. Soc., 1958:4382.

146. M. H. Kaufman, F. M. Ernsberger, and W. S. McEvan, J. Am. Chem. Soc., 78: 4197 (1956).

147. C. P. Smyth, Dielectric Behavior and Structure, McGraw-Hill, New York (1955), Chap. 10.

148. E. A. Shott-L'vova, Ya. K. Syrkin, I. I. Levkoev, and M. B. Deichmeister, Dokl. Akad. Nauk SSSR, 121:1048 (1959).

149. A. I. Kiprianov, Usp. Khim., 29:1336 (1960).

150. I. Yu. Kokoreva, L. A. Neiman, Ya. K. Syrkin, and S. I. Kirillova, Dokl. Akad. Nauk SSSR, 156:412 (1964).

151. A. W. Sharpe and S. Walker, J. Chem. Soc., 1961:4522.

152. Z. V. Pushkareva, L. V. Varyukhina, and Z. Yu. Kokoshko, Dokl. Akad. Nauk SSSR, 108:1098 (1956).

153. E. N. Gur'yanova, Zh. Fiz. Khim., 24:479 (1950).

154. E. N. Gur'yanova, I. P. Gol'dshtein, E. N. Prilezhaev, and L. P. Tsimbal, Izv. Akad. Nauk SSSR, Ser. Khim., 810 (1962).

155. K. A. Jensen, Z. Anorg. Chem., 250:268 (1943).

156. C. P. Smyth, Dielectric Behavior and Structure, McGraw-Hill, New York (1955), Chap. 8.

157. C. W. N. Cumper, A. A. Foxton, J. Read, and A. I. Vogel, J. Chem. Soc., 1964: 430.

158. B. A. Arbousow, Bull. Soc. Chim. France, 1960:1311

159. H. B. Thomson, L. P. Kuhn, and M. Inatome, J. Phys. Chem., 68:421 (1964).

160. H. C. Fu, T. Psarras, H. Weidmann, and H. H. Zimmerman, Ann. Chem., 641: 116 (1961).

161. A. T. Balaban, J. Bally, R. J. Bishop, C. N. Reuten, and L. E. Sutton, J. Chem. Soc., 1964:2383.

162. I. A. Sheka, Papers on the Chemistry of Solutions and Complex Compounds, Izd. Akad. Nauk SSSR (1954), p. 73.

163. O. A. Osipov, V. I. Minkin, and V. A. Kogan, Zh. Fiz. Khim., 36:889 (1963).

164. I. P. Gol'dshtein, E. I. Gur'yanova, and K. A. Kocheshkov, Dokl. Akad. Nauk SSSR, 144:569 (1962).

165. P. A. McCuscer and S. M. L. Kilzer, J. Am. Chem. Soc., 82:372 (1960).

166. A. D. Garnovskii, O. A. Osipov, and V. I. Minkin, Usp. Khim. (1968).

167. O. A. Osipov, A. D. Garnovskii, and V. I. Minkin, Zh. Strukt. Khim., 8:919 (1967).

168. N. J. Leonard, Rec. Chem. Progr., 17:243 (1956).

169. L. N. Ferguson and J. C. Nnadi, J. Chem. Educ., 42:529 (1965).

170. N. J. Leonard, D. F. Morrow, and M. T. Rogers, J. Am. Chem. Soc., 79:5476 (1957).

171. N. J. Leonard and C. R. Johnson, J. Am. Chem. Soc., 84:3701 (1962).

172. J. M. Pugh and R. H. Stokes, Australian J. Chem., 16:204 (1963).

173. M. G. Voronkov, I. B. Mazheika, and G. I. Zelchan, Khim. Geterotsikl. Soedin, 1:58 (1965).

174. M. G. Voronkov, Pure Appl. Chem., 13:35 (1966).

175. E. A. Shott-L'vova, Ya. K. Syrkin, I. I. Levkoev, and Z. P. Sytnik, Dokl. Akad. Nauk SSSR, 116:804 (1957).

176. M. H. Hutchinson and L. E. Sutton, J. Chem. Soc., 1958:4382.

Chapter VI

Dipole Moments and Some Special Problems of the Structure and Properties of Organic Compounds

In this chapter we shall discuss the application of the dipole moment method to the investigation of a number of typical problems arising in the study of the structure and properties of organic compounds in solutions. Particular attention is devoted to the characteristics of the methods of determination and to the significance of the information provided by a study of the dipole moments of molecules or organic compounds in electronically excited states.

Tautomerism

If two tautomeric forms that exist or are expected in a solution of any compound differ considerably in the magnitudes of the dipole moments predicted for them, a comparison of the latter with the experimental value can give useful information on the position of the tautomeric equilibrium. In principle, also, an experimental value of the dipole moment intermediate between those calculated for the two tautomers can be used to evaluate the constant of the tautomeric equilibrium, but in practice it is rare to encounter cases in which the structure of each of the tautomers is not complicated by subsidiary factors (internal rotation, supplementary conjugation effects, etc.) which permit only an extremely approximate evaluation of the theoretical values. In contrast to such cases of the investigation of equilibria by means of dipole moments as, for example, conformation equilibria (Chapter IV) (where the moments of the individual forms can be established by determining the dipole moments of model compounds), in a study of tautomerism it is impossible to select equally suitable models, since the difference in the dipole moments of the fixed alkyl analogs of the individual tautomeric structures and of the tautomers themselves may amount to 1 D. In view of this, the dipole moment method is used mainly for potentially tautomeric compounds which exist in solutions exclusively or predominantly in the form of one of the tautomers. In this case, the problem reduces to determining which of the possible tautomers does in fact predominate.

A characteristic example is the study of the thiol-thione tautomerism of 2-mercaptobenzothiazole [1,2]

The dipole moment expected for the thione structure (I) on the basis of vector calculations is 4.4-4.6 D, while the moment of the thiol (II) should not exceed 2-2.4 D. The experimental value obtained in measurements in benzene at 25°C is 4.67 D and puts the predominance of the thione form (I) beyond doubt. At the same time, when the temperature is raised the dipole moment of 2-mercaptobenzothiazole is found to fall considerably, indicating the gradual shift of the prototropic equilibrium in the direction of the less polar thiol structure (II).

Similarly, it has been possible to show by means of dipole moments the complete or very pronounced displacement of the amine-imine tautomerism of the 2-aminopyridines in the direction of the amino form [3], the predominance of the hydrazone form of the phenylhydrazones [4], of the hydroxy azo form of the o-hydroxyazobenzenes [5] and the o-hydroxyphenylazoazulenes [6], of the monoenol forms of 1,2-cyclopentane- and 1,2-cyclohexanediones [7], and the energetic advantage of the benzoid structure of the salicylidenearylamines (III) over the quinoid structure (IV) [8, 9]:

R	μ_{calc}, D	μ_{exp}, D	μ_{calc}, D
H	2.7	2.39	2.8
NO$_2$	4.4	4.03	1.2
Cl	2.3	2.16	1.3

The last example is interesting because the dipole moments calculated for the two tautomeric forms of the unsubstituted sali-cylideneaniline almost coincide; however, the choice between structures (III) and (IV) is made on the basis of the moments of anils substituted in the aldehyde-containing nucleus.*

It is particularly easy to establish the nature of the predom-inating tautomeric structure in solution if one of the tautomers is a bipolar ion as, for example, in the case of 3-hydroxypyridine

The dipole moment of 3-hydroxypyridine in benzene is 2 D, which shows the absence of appreciable amounts of structure (VI), the moment of which should be about 16 D in a nonpolar solvent [10].

As already mentioned, a comparison of the dipole moments of a tautomeric compound and the alkyl analogs of each of the taut-omers cannot be regarded as such a direct method of studying a tautomeric equilibrium as, for example, a comparison of the spec-tra of these compounds. However, if the expected moments of the tautomers differ by not less than 2-3 D, such a comparison may be a fairly reliable means of determining the more stable tautomer. Thus, the dipole moment of 4-hydroxypyridine in benzene is 6.0 D [11], and in dioxane it is 6.3 D [10]. A comparison of these figures with the moments of the fixed O- and N-alkyl analogs of the tauto-meric forms (VII) and (VIII)

---

* Spectroscopic studies have shown that the introduction of substituents into the 5 posi-tion of the aldehyde-containing nucleus has no effect on the position of the equili-brium, but they give no clear information on which form predominates in solution.

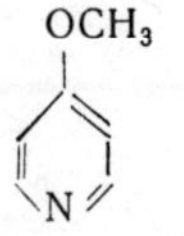

$\mu = 2.94\,D$ [11]

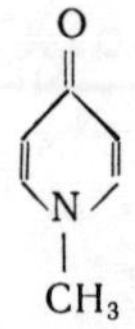

$\mu = 6.9\,D$ [10]

shows quite clearly that there is a displacement of the tautomeric equilibrium of 4-hydroxypyridine in solutions in the direction of the pyridine form (X).

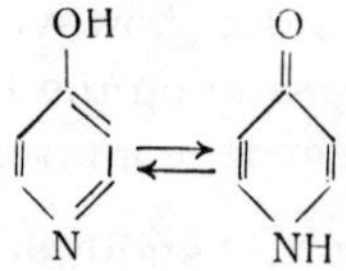

This result is confirmed by the results of other methods of investigation [10].

The dipole moment method has also been used to study the tautomeric equilibria of triazenes [12] and of thioacetic ester [13] and the ring-chain tautomerism of the enamines [14] and of the dibromopyrazoles [15].

2. The Hydrogen Bond

Dipole moments give important information on the nature and characteristics of such a peculiar type of chemical interaction as the hydrogen bond.

The Polarity and Nature of the Hydrogen Bond. Even on the basis of a simple comparison of the dipole moments of nitromethane (3.10 D) and acetonitrile (3.47 D), on the one hand, and dioxane (0-0.4 D) and acetone (2.8 D), on the other hand, and also of spectroscopic data showing that the first group of bases forms less stable complexes through hydrogen bonds (H complexes) with phenols and other acids than the second group, it is clear that the hydrogen bond cannot have a purely electrostatic nature. Modern ideas on the nature of the hydrogen bond (see reviews [16-20]), the basic starting points of which were first developed by Sokolov [2, 21], involve a consideration of the wave function of a H complex between the hydrogen donor AH and its acceptor B in the form of a superposition of at least the following three valence structures:

I. $A-H : B$ — covalent
II. $A^- H^+ : B$ — ionic without charge transfer
III. $A^- H - B^+$ — ionic with charge transfer

The contribution of structures (II) and (III) to the complete wave function of the H complex must lead to an increase in its polarity as compared with the additive value. It must also be expected that an increase in the dipole moment of the H complex through the hydrogen bond will depend on the properties of its components. The data given in Table 50, which have been collected from various sources, strikingly confirm these predictions. The dipole moments of a H complex (μ_{compl}) partially dissociated into its components even in a nonpolar solvent were determined by one of the methods considered below (Section 3 of this chapter). The penultimate column of Table 50 gives the magnitudes $\Delta\mu = \mu_{compl} - \mu_{AH} - \mu_B$ characterizing the polarity of the hydrogen bond.*

As can be seen from the data of Table 50, $\Delta\mu$ rises with an increase in the acidity of AH (fall in pK_a). Reference to this nature of the dependence of the polarity of the hydrogen bond on the properties of the hydrogen donor was first made by Sobczyk and Syrkin [29] and found subsequent confirmation in a whole series of investigations [23-26]. There is less detailed information on the question of the influence on the polarity of the hydrogen bond of the basicity of the hydrogen acceptor. However, a comparison of the data of Table 50 for the complexes of the trialkylamines and pyridine with phenols shows that the values of $\Delta\mu$ for the H complexes are considerably smaller than for pyridine base and far lower than in the case of the trialkylamines.

An analysis of the dipole moments of the H complexes enables interesting conclusions to be drawn about the mechanism of the acid-base interaction in nonpolar solvents and the nature of the

---

* It is possibly more correct to consider the vectorial difference, taking into account the fact that the vector of the moment contributed by the hydrogen bond is directed along the hydrogen bond and has its positive pole on B [23, 27, 28]. This method, however, also does not take into account the electronic redistributions in other parts of the H complex and requires assumptions concerning its geometry.

TABLE 50. Dipole Moments of H Complexes AH $\cdots$ B in Benzene

B	AH	pK_aAH	μ_{AH}, D	μ_{compl}, D	$\Delta\mu$, D	Literature
Triethylamine	Butanol	19	1.66	2.24	−0.28	[25]
($\mu = 0.86$ D)	p-Cresol	10.17	1.55	2.73	0.32	[23]
	Phenol	9.95	1.59	2.96	0.51	[23]
	p-Nitrophenol	7.15	5.06	7.26	1.34	[23]
	Acetic acid	4.75	1.68	4.02	1.48	[25, 26]
	Benzoic acid	4.20	1.61	4.38	1.91	[25, 26]
	Chloroacetic acid	2.86	2.29	6.80	3.65	[25, 26]
	Picric acid	0.70	4.76	11.66	9.28	[25, 26]
Tributylamine	Benzoic acid	4.20	1.61	4.24	1.75	[25, 26]
($\mu = 0.88$ D)	p-Fluorobenzoic acid	3.69	1.83	7.38	4.67	[25, 26]
	p-Nitrobenzoic acid	3.44	3.12	8.98	4.98	[25, 25]
	2,4,6-Tribromo-benzoic acid	1.41	1.81	6.93	4.24	[25, 26]
	2,4,6-Trinitro-benzoic acid	0.65	1.95	8.54	5.71	[25,26]
	Hydrobromic acid	—	1.08	8.50	6.54	[25, 26]
Pyridine	Butanol	19	1.66	3.37	−0.51	[25]
($\mu = 2.22$ D)	p-Cresol	10.17	1.55	3.66	−0.11	[23]
	Phenol	9.95	1.59	3.86	0.05	[23]
	p-Chlorophenol	9.42	2.19	4.57	0.16	[23]
	p-Nitrophenol	7.15	5.06	7.23	−0.05	[23]
	Trichloroacetic acid	0.63	1.90	7.78	3.66	[28]

potential curve of the hydrogen bond in the A $-$H$\cdots$B systems [23, 25, 26].

The dipole moments of the H complexes of the trialkylamines with strong acids such as picric and hydrobromic acids (Table 50) are so large that we are forced to assume the presence in solutions of these H complexes of a considerable amount of the ion-pair form (V)

$$A-H+B \rightleftarrows \underset{IV}{A-H\cdots B} \rightleftarrows \underset{V}{A^-\cdots H-B^+} \qquad (VI.1)$$

This assumption is confirmed by data on the dipole moments of the tetrabutylammonium salts (VI) with the corresponding anions, which exist in benzene solutions in the form of ion pairs

$$
\begin{array}{ccc}
C_4H_9 \\
|_+ \\
C_4H_9-N-C_4H_9 \quad An^- \\
| \\
C_4H_9 \\
VI
\end{array}
\qquad
\begin{array}{l}
An^- \\
Br^- \\
C_6H_2(NO_2)_3O^- \\
CH_2ClCOO^- \\
C_6H_5COO^- \\
CH_3COO^-
\end{array}
\qquad
\begin{array}{l}
\mu, D \\
12.2\ [25,\ 26] \\
15.3\ [25,\ 26,\ 30] \\
14.8\ [25,\ 26] \\
12.1\ [25,\ 26] \\
11.2\ [5]
\end{array}
$$

Although the dipole moments of tetrabutylammonium picrate and bromide are approximately 3.5 D higher than those of the corresponding H complexes, it must be borne in mind that this may be due to a decrease in the distance between the cation and the anion in (V) through the formation of the hydrogen bond, leading to a decrease in the interionic distance. On the other hand, the dipole moments of tetrabutylammonium chloroacetate, benzoate, and acetate are higher than the moments of the corresponding H complexes by 7–8 D. Consequently, the content of ion pairs of type (V) in the tautomeric equilibrium (VI.1) for such acids, in contrast to stronger ones, is comparatively low. Finally, for weak acids, the complex equilibrium (VI.1) is limited to the first stage and is not accompanied by a detectable formation of ion pairs.

The advantages of the dipole moment method for studying the possibility of the existence of ion pairs in solutions may also be illustrated by the following example. On the basis of an interpretation of the IR spectra of imidazole, Otting [31] came to the conclusion that imidazole and its derivatives exist in solutions in the form of ion pairs produced by the transfer of the proton from one molecule to another, for example:

$$
2
\begin{array}{c}
HC-N \\
\| \quad \ \ \ CH \\
HC-NH
\end{array}
\ \rightleftharpoons\
\left[
\begin{array}{c}
HC-N \\
\| \ \ominus\ CH \\
HC-N
\end{array}
\right]
\left[
\begin{array}{c}
HC-NH \\
\| \ \oplus\ CH \\
HC-NH
\end{array}
\right]
\qquad (VI.2)
$$

However, a determination of the dipole moments of a wide series of imidazole derivatives has shown that their magnitudes (about 4 D) are not large enough for it to be possible to connect them with the presence of the equilibrium (VI.2) displaced to the right. Moreover, the moments of the N-methyl derivatives, for which equilibria of type (VI.2) are unrealizable, scarcely differ from the moments of the NH derivatives [32, 33].

TABLE 51. Dipole Moments and the Dioxane Effect

Compound	μ, D (benzene)	μ, D (dioxane)	$\Delta\mu = \mu_{dioxane} - \mu_{benzene}$	Literature
Aniline	1.53	1.77	0.24	[25]
p-Aminobenzonitrile	5.96	6.45	0.49	[25]
p-Nitroaniline	6.22	6.82	0.60	[25]
2-Amino-6-nitronaphthalene	6.04	7.13	1.09	[41]
Phenol	1.56	1.86	0.30	[39]
2-Naphthol	1.58	2.02	0.44	[34]
1-Nitro-4-naphthol	5.27	5.88	0.61	[34]
Palmitic acid	0.76	1.75	0.99	[39]
Salicylaldehyde	2.91	3.02	0.11	[39]
o-Hydroxyacetophenone	3.19	3.26	0.07	[39]
Guaiacol	2.37	2.40	0.03	[39]
Benzylidene-o-aminophenol	2.73	2.68	-0.05	[42]
2-Acetyl-1-naphthol	3.14	3.12	-0.02	[34]
1-Nitro-2-naphthol	3.91	4.07	0.16	[34]
8-Hydroxyquinoline	2.68	2.51	-0.17	[39]
o-Nitroaniline	4.38	4.68	0.30	[34]
o-Nitro-N-methyldiphenylamine	3.64	3.81	0.17	[34]
Dimethylaniline	1.61	1.66	0.05	[36]
Anisole	1.33	1.36	0.03	[39]
p-Nitrodimethylaniline	6.96	7.27	0.31	[40]
Dimethylaniline N-oxide	4.79	4.85	0.06	[39]
1-Nitro-2-methoxynaphthalene	4.66	4.86	0.20	[34]
1-Nitro-4-methoxynaphthalene	5.19	5.27	0.08	[34]

   Intramolecular Hydrogen Bond. If the A—H group
and B are present in one and the same molecule and are sufficiently close to one another ($l_{A...B} = 2.4$-3.0 Å), an intramolecular hydrogen bond may arise. The tasks of the dipole moment method in a
study of compounds for which intramolecular hydrogen bonds are
possible reduce to 1) a determination of the presence or absence
of an intramolecular hydrogen bond in the molecule; 2) characterizing its strength; and 3) evaluating its influence on the distribution of the electron density in the molecule.

   The dipole moments of aromatic compounds with 6- and 5-
membered rings formed by intramolecular hydrogen bonds of type
(VII) have been studied most systematically.

$$R \underset{\substack{B}}{\overset{\substack{A}}{}} H \qquad \begin{array}{l} A = O, \ NH, \ NAlk, \ NAr, \ NCOCH_3 \\ B = CHO, \ COCH_3, \ NO_2, \ NO, \ OAlk, \ NH_2 \end{array}$$

VII

   In a number of papers, Lutskii et al. [34] have formulated
the main empirical criteria for the presence of intramolecular hydrogen bonds in molecules resulting from a study of dipole moments: 1) an anomalous (see Table 50) reduction (generally by
1-2 D) of the dipole moment of a compound (VII) with an intramolecular hydrogen bond as compared with the moment of its para
isomer, in which an intramolecular hydrogen bond cannot be formed
for steric reasons; 2) a reduction of the dipole moment as compared with that calculated from formula (III.10) for nonfixed orientations of A and B; 3) extremely close agreement of the dipole
moments of a compound with an intramolecular hydrogen bond and
a compound of type Ar—B not containing the A—H group; 4) a considerable (generally about 1 D) increase in the dipole moment of
the methyl ethers of compounds of type (VII) as compared with the
initial substances while the opposite situation is characteristic for
the para isomers; and 5) a small "dioxane effect," i.e., a small
difference in the magnitude of the dipole moments determined in
dioxane and in benzene.

   The last item is particularly important, and besides Lutskii's
papers [34], a number of others [35-39] have been devoted to it. The
dioxane effect can be used as a test to determine the intramolecular or intermolecular nature of a hydrogen bond. Since dioxane is
a fairly strong base, it is capable of forming H complexes with

compounds having acidic OH, NH, etc. groups. The additional polarity of the hydrogen bond leads to the situation that in the calculation of the dipole moment for the molecule of a dissolved substance the latter rises considerably as compared with the figure
determined in a solution not capable of forming intermolecular
hydrogen bonds. Another, possibly more important, cause of the
dioxane effect is the limitation in the H complex of the internal rotation of the group A —H and electron redistribution, leading to an
increase in the observed dipole moment [19, 25].

Both the causes mentioned lose their force for compounds in
which the acid group A —H is absent or takes part in a string intramolecular hydrogen bond which is not disturbed by interaction with
dioxane. Consequently, for such compounds the dioxane effect
must be absent and the differences in the dipole moments measured
in benzene and in dioxane will be determined only by the ordinary
universal interaction with the solvent.

A confirmation of what has been said is formed by the data
given in Table 51. The compounds given in it are divided into
three groups. The first includes compounds tending to form intermolecular hydrogen bonds with dioxane, the second compounds
with strong intramolecular hydrogen bonds, and the third some
compounds the structure of which excludes both the existence of
intramolecular hydrogen bonds and the formation of H complexes
with dioxane.

In molecules with π-electrons in which the functional groups
AH and B form part of a conjugated system, the hydrogen bond is
characterized by a number of special features [43]. The most important of them is the capacity of the hydrogen bond in such molecules for transferring conjugation effects, i.e., the delocalization
of the π-electrons through a hydrogen bond [44]. This circumstance
plays a fundamental role in a number of vitally important biologi-
cal processes [45, 46].

The delocalization of the π-electrons through a hydrogen
bond, if it takes place, obviously leads to the appearance of an in-
crement in the dipole moment, the vector of which $\Delta\vec{\mu}_H$ can be determined by comparing the experimental value of the dipole mo-
ment $\vec{\mu}_{exp}$ with the magnitude $\vec{\mu}_{calc}$, without taking the perturbation
caused by the hydrogen bond into account, $\Delta\vec{\mu}_H = \vec{\mu}_{exp} - \vec{\mu}_{calc}$.

The scheme of calculating the magnitude of $\overrightarrow{\Delta\mu_H}$ was proposed by Eda and Ito [47], who interpreted the dipole moments of salicylaldehyde and o-nitrophenol as those of compounds with six-membered rings formed by intramolecular hydrogen bonds. Let us consider it on the basis of another example, the azomethines (VIII) and (IX) with, respectively, six- and five-membered rings formed by intramolecular hydrogen bonds [42, 48].

A—$m_x = -1.34$ (here and below, all the values of μ and their projections are given in Debyes); $m_y = 0.78$; B —$m_x = 0.84$; $m_y = -0.95$; C —$m_x = 1.34$; $m_y = 0.55$; $m_z = -0.55$; D—$m_x = -0.84$; $m_y = -0.47$; $m_z = -0.82$; E — $m_x = -0.84$; $m_y = 0.47$; $m_z = -0.82$.

In the given system, the vector $\overrightarrow{\mu}_{calc}$ of salicylideneaniline (VIII, $R_1 = R_2 = H$) can be represented as the sum of the vectors of the moments of benzylideneaniline and phenol in configuration A, and $\overrightarrow{\mu}_{calc}$ for benzylidene-o-aminophenol (IX, $R = C_6H_5$) as the sum of the vectors of the moments of benzylideneaniline and phenol in configuration C. The vector of the moment of the phenol (like that of anisole and p-bromophenol) is known (Table 10) and needs only to be transformed according to the system of coordinates, and the vector of the moment of benzylideneaniline is calculated from $\overrightarrow{\mu}_{exp}$ for benzylideneaniline (X), benzylidene-p-toluidine, and the moment of the CH_3 group (Chapter III, Section 3).

The values of $\overrightarrow{\mu}$ for other anils the molecules of which serve as frameworks for compounds with an intramolecular hydrogen bond can be determined similarly.

Table 52 gives the values of $\overrightarrow{\mu}_{calc}$ calculated by the method given above (without taking into account the new distribution of the

TABLE 52. Influence of a Hydrogen Bond on the Dipole
Moments (in D) of the Azomethines (VIII)

Compound	$\vec{\mu}_{calc}$		μ_{calc}	$\vec{\mu}_{exp}$		μ_{exp}	$\overrightarrow{\Delta\mu_H}$		$\Delta\mu_H$
	m_x	m_y		m_x	m_y		m_x	m_y	
$R_1 = R_2 = H$	—1.34	2.38	2.73	—1.44	1.91	2.39	—0.10	—1.47	0.47
$R_1 = 5\text{-Br}, R_2 = H$	—2.12	1.02	2.36	—2.16	0.16	2.16	—0.04	—0.86	0.86
$R_1 = 4,5\text{-}C_6H_4, R_2 = H$	—1.01	2.41	2.62	—1.42	1.60	2.18	—0.41	—0.81	0.90

electrons due to the formation of a ring by the hydrogen bond) and
$\vec{\mu}_{exp}$, in which this effect is taken into account. The vector $\vec{\mu}_{exp}$
was determined from the moments of compounds (VIII) with $R_2 = H$
and $R_2 = p\text{-}CH_3$. The last column gives the vector $\vec{\Delta\mu}$. The figures
of this table show that the appearance of a hydrogen bond leads to
a decrease in the dipole moment as compared with the additive
value μ_{calc} by 0.5–0.9 D. At the same time, the vector of the additional moment $\vec{\Delta\mu}$ has negative x- and y-components, i.e., its positive pole is located on the oxygen atom and not on the nitrogen. In
order to be able to state with certainty that the appearance of
is a consequence of a hydrogen bond, we must establish that the
difference between $\vec{\mu}_{exp}$ and $\vec{\mu}_{calc}$ is not simply the result of an increase in the conjugation of the p-electrons of the hydroxyl with
the π-electrons of the aromatic nucleus attached to the electron-
accepting aryl azomethine group. If there is no such additional
conjugation, $\vec{\mu}_{calc}$ and $\vec{\mu}_{exp}$ must agree for the methyl ethers (VIII)
and (IX). In actual fact, the agreement of these values is complete-
ly satisfactory and so is the agreement of the calculated and experimental values in the case of the azomethines (IX) with five-
membered rings formed by hydrogen bonds (Table 53).

The difference between $\vec{\mu}_{exp}$ and $\vec{\mu}_{calc}$ for the azomethines
(VIII) cannot be explained as a contribution of ionic structures of
type (III) or as a result of the intramolecular transfer of a proton
of type (V). On the other hand, calculation by the LCAO MO method [48] on the assumption of delocalization through the hydrogen
bond enables the values of $\vec{\Delta\mu}_H$ in Table 52 to be reproduced correctly, which serves as a confirmation of the correctness of the

TABLE 53. Dipole Moments (in D) of the
Azomethines (IX) (Benzene, 25°C)

R	μ_{exp}	μ_{calc}
Phenyl	2.73	2.60
p-Tolyl	3.20	3.26
p-Dimethylaminophenyl	5.08	5.02
9-Anthryl	2.79	2.92
Ferrocenyl	3.69	3.68

ideas on the capacity of hydrogen bonds for transferring conjugation effects.

At the same time, as can be seen from the data of Tables 52
and 53, the properties of intramolecular hydrogen bonds in six-
and five-membered rings must be delimited. In the latter case
[compounds of type (IX)] the existence of a strong intramolecular
hydrogen bond is not connected with such a fundamental electronic
redistribution as in compounds (VIII) with six-membered hydrogen
bond rings.

The nature of some compounds implies the possibility of re-
alizing not one but two structures with intramolecular hydrogen
bonds as is the case, for example, for 2-(o-hydroxyphenyl)benzox-
azole:

XI   XII
$\mu_{calc} = 2.38\ D$ $\mu_{calc} = 1.36\ D$

In this case, the choice can be made on the basis of a comparison
of the experimental value of the dipole moment with those calculated for each structure. The dipole moment of 2-(o-hydroxyphenyl)-
benzoxazole is 2.10 D in benzene and 2.15 D in dioxane [49]. A
comparison with the values calculated by the method given above
on the basis of the azomethines clearly shows the formation of a
ring by the hydrogen bond with the nitrogen atom — structure (XI)
[49].

A consideration of the material given in the present section
permits the conclusion that in the study of a number of aspects of

the problem of the hydrogen bond the dipole moment method gives extremely interesting and specific information which cannot be obtained by other methods in many cases.

3.   Intermolecular Interactions

An example of the investigation of intermolecular interactions by the dipole moment method is the above-described study of the H-complexes formed in solutions containing bases and compounds with labile hydrogen atoms. In the present section we shall briefly consider other types of intermolecular interactions, information on the nature of which and on the structure of the corresponding associates is given by measurements of polarization and dipole moments.

Calculation of the Reaction Constants for Complex Formation and of the Dipole Moments of the Complexes. If the compound, the dielectric constant of a solution of which is measured, is capable of self-association in solution or of chemical interaction with the solvent or if two components are present in solution which take part in a complex-forming reaction, the experimentally determined molar polarization is composed additively of the polarizations of all the forms existing in the solution: the solvent, the monomer, the dimer, the associate, etc. In connection with this, problems arise of establishing the position of the equilibrium between the forms present in the solution and finding the dipole moment of the complex produced. As a rule, the composition of such a complex is known from data obtained by any other method, but in principle it can also be determined on the basis of a study of the dependence of the dielectric constant on the ratio of the components in the solution [50–52].

For the case of a dissociating complex of composition AB (1 : 1) present in solution in a nonpolar solvent S, the method proposed by Smith et al. [53–55] is particularly suitable. To find the constant of the equilibrium established in solution: $A + B \rightleftarrows AB$, and to calculate the dipole moment of the complex AB we must determine the apparent molar polarization of one of the components A in a series of solutions of B in S of constant concentration. To calculate K_{AB} and μ_{AB}, the following relation is used:

$$\frac{1}{(P_{A\infty})_{BS} - (P_{A\infty})_S} = \frac{M_B}{w_B d_{BS}} \cdot \frac{1}{K_{AB} \Delta P} + \frac{1}{\Delta P} \qquad (VI.3)$$

where $(P_{A\infty})_{BS}$ and $(P_{A\infty})_S$ are the molar polarizations of component A in mixtures of B and S and in the solvent S, respectively, obtained by extrapolation to zero concentration; M_B, w_B, and d_{BS} are the molecular weight, weight fraction, and density of component B and its solutions in the nonpolar solvent S; and $\Delta P = P_{AB} - P_A - P_B$.

Relation (VI.3) is a general linear function, and plotting $1/[(P_{A\infty})_{BS} - (P_{A\infty})_S]$ versus $M_B/w_B d_{BS}$ gives a straight line the angle of slope of which to the axis of abscissas is $1/K_{AB}\Delta P$, while the intercept cut off on the axis of ordinates is $1/\Delta P$. Knowing ΔP it is easy to calculate μ_{AB} and K_{AB}.

Smith's method has been used particularly widely for determining the equilibrium constants and dipole moments of H complexes (see Table 50); however, the nature of the forces of interaction between A and B leading to complex formation do not impose any limits on the applicability of relation (VI.3) to the calculation of K_{AB} and μ_{AB}. Smith [25] has shown that the equilibrium constants K_{AB} found according to (VI.3) agree well with the results obtained by spectroscopic methods.

Bishop and Sutton [56], performing measurements at several temperatures and calculating K_{AB} from (VI.3), were able to determine with acceptable accuracy the thermodynamic parameters of the reaction of complex formation through a hydrogen bond. They showed that when the solvent (for example, benzene) is itself capable of forming associates with the components A and B the true equilibrium constant K'_{AB} can be derived from K_{AB} (VI.3) by means of the relation

$$K'_{AB} = K_{AB} (1 + c K_{AS})(1 + c K_{BS}) \qquad (IV.4)$$

where K_{AS} and K_{BS} are the association constants of the components A and B with the solvent and c is the concentration of the solvent.

As already mentioned, the method expounded above can be used only to calculate equilibrium constants and dipole moments of complexes with the 1 : 1 composition. Osipov and Shelomov [57, 58] have developed a method of determining the complex-formation constants and the dipole moments of complexes of any composition. The essence of the method consists in determining the molar polarization of a complex partially dissociated in solution in an excess of one of the components capable of suppressing its dissociation. For this purpose, the molar polarization is extrapolated to infinite excess of the component selected.

Another method enabling the dipole moments of complexes of any composition to be determined has been proposed by Gur'yanova and Gol'dshtein [59]. They studied the change in the dielectric constant ε and the density d of solutions of component A in a nonpolar solvent with the successive addition of small portions c of the second component B. Curves of the dielectrometric titration in terms of $\triangle\varepsilon/c_B$ versus c_A/c_B show characteristic maxima or breaks which enable the composition of the complexes to be followed. We may also mention other methods of calculating dipole moments and the composition of complexes [60, 61].

Dipole Association. This is one of the most universal methods of intermolecular interaction. When two polar molecules with the dipole moments μ_1 and μ_2 approach one another in solution in a nonpolar solvent, forces of dipole—dipole attraction arise which are inversely proportional to the cube of the distance R between the dipoles and the dielectric constant of the medium ε [62, 63]:

$$U_\mu = \frac{\mu_1\mu_2}{\varepsilon R^3}(\cos\alpha - 3\cos\beta_1\cos\beta_2) \tag{VI.5}$$

Here the term included in brackets characterizes the relative orientation of the axes of the dipoles.

The dimer arising as a result of the mutual attraction of two molecular dipoles will be stable if the energy of the dipole attraction U_μ is appreciably greater than kT. For two point dipoles with moments of 5 D approaching to a distance of 5 Å in a medium with $\varepsilon = 2$, the ratio $U_\mu/kT = 5$. Consequently, for highly polar molecules fairly stable dimers may be expected to exist in solutions in nonpolar solvents. Depending on the geometry of the molecule, the stable configuration of the dimer will be either that with the parallel or that with the antiparallel arrangement of the axes of the dipoles. In the first case, the volume polarization will increase with a rise in the concentration of the polar substance far more rapidly and in the second case more slowly than would be found in the absence of an interaction between the polar molecules.

Treiner, Skinner, and Fuoss [64] have recently proposed an extremely convenient method for treating ordinary experimental data (the dependence of the dielectric constant and the density of solutions on the concentration) which makes it possible to calculate not only the dipole moment of a monomer but also its dimerization constant K_d. The values of K_d (in liter/mole) for m-nitrophenol (I), p-nitroaniline (II), and pyridinium dicyanomethylide (III) in dioxane are, respectively, 0.37, 0.8, and 3:

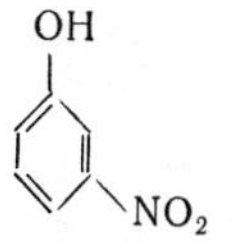

I

$\mu = 4.38\ D$

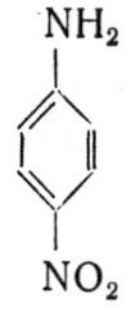

II

$\mu = 6.91\ D$

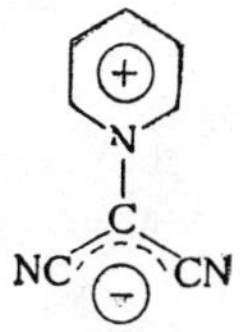

III

$\mu = 9.2\ D$

In agreement with theoretical expectations, the logarithms of these constants are proportional to the squares of the dipole moments of the monomers:

$$K \sim e^{-\frac{U_\mu}{kT}}\ ;\quad U_\mu \sim \mu^2$$

<u>Association through Hydrogen Bonds</u>. The nature of this effect has been considered above (Section 2). Information on the existence of associates in solution can be obtained by studying the dependence of the apparent molar polarization of a substance (calculated to the monomer) on its concentration. In the majority of cases, it is fairly simple to follow the concentration dependence of the dielectric constant: considerable deviation from linearity definitely indicates association [59, 65, 66].

A characteristic example of the formation of associates in solution through intermolecular hydrogen bonds is that of imidazole derivatives. At sufficiently high concentrations, the association coefficient of imidazole itself (R = H) in nonaqueous solutions amounts to 5–20 [67]

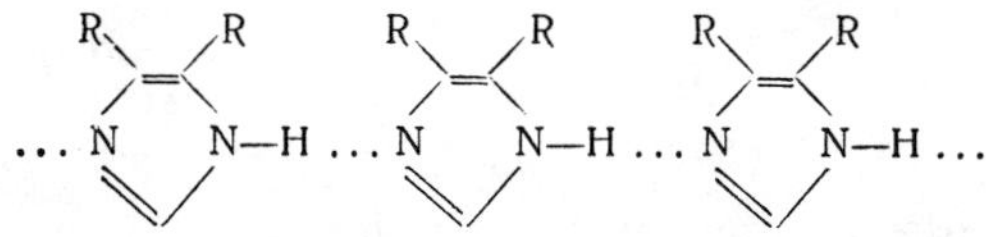

IV

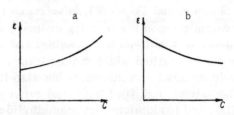

Fig. 25. Types of relationships between the dielectric constant and the concentration of solutions of an associated compound in a nonpolar solvent: a) associate more polar than the monomer; b) associate less polar than the monomer.

Because of the presence of hydrogen bonds, the vectors of the moments of which are collinear in this associate, the associate possesses a considerably higher polarity than the monomer, and the graph of the dielectric constant as a function of the concentration has the form shown in Fig. 25a.

In view of this, the dipole moments of imidazole and its derivatives based on measurements of concentrated solutions obtained in a number of studies [68, 69] are 2-2.5 D higher than the data of later studies [32, 33, 71] in which extrapolation was carried out to infinite dilution. It is interesting that in the case of 4,5-diphenylimidazole ($R = C_6H_5$), the relationship $\varepsilon = f(c)$ is rectilinear: there is no association. The reason for this is obviously the steric hindrance created by the voluminous phenyl groups.

If association leads to the formation of an associate less polar than the monomer, the relation $\varepsilon = f(c)$ is similar to that shown in Fig. 25b. This case is most characteristic when an equilibrium between the monomer and its less polar dimer exists in solution as, for example, for the pyrrolidone (V) [73-75], and some other cyclic lactams [74]

The dipole moment of 2.3 D calculated in an earlier paper [72] is too small in view of the fact that the presence of the feebly

polar dimer (VI) in solution was not taken into account. The true
value of the dipole moment of (V) is 3.79 D (dioxane) [73]. For
comparison, we may mention that the N-methyl derivative of (V)
has a dipole moment of 4.06 D [73].

Charge Transfer Complexes.* A particularly interesting example of intermolecular interaction due to the appearance of specific forces of chemical affinity is the formation in solution of so-called charge transfer complexes (CTC). In compounds
of this type, a definite part of the energy of association is due to
the intermolecular transfer of charge from the donor component
of the complex D to the acceptor component A, the degree of this
being characterized by the contribution of the ionic term $b\psi_1$ to the
total wave function of the ground state of the complex [76]:

$$\Psi_N = a\Psi_0\,(A,\ D) + b\Psi_1\,(A^-,\ D^+) \qquad\qquad (VI.6)$$

It is obvious that the dipole moment, a magnitude determined by
the electrical asymmetry of the molecule, can give valuable information on the role and magnitude of the transfer of charge in a
CTC. It can be used for an approximate evaluation of the coefficients a and b in the wave function (VI.6). For a complex formed by
a nonpolar donor and a nonpolar acceptor, or by a donor and acceptor having small dipole moments, the following relation is satisfied [76, 78]:

$$\mu_{\exp} = \mu_1\,(b^2 + abS) \qquad\qquad (VI.7)$$

where μ_1 is the dipole moment of the structure ψ_1 with full charge
transfer and S is the overlap integral between the functions ψ_0 and
ψ_1, which is readily reducible to an integral with respect to the
atomic orbitals.

Depending on the type of donor center, it is convenient to
isolate two types of CTC: p-complexes and π-complexes. In the
first of them, the transfer of charge to the orbitals of the acceptor
is effected from nonbinding orbitals of oxygen, nitrogen, or sulfur
atoms present in the donor molecule. The first indication of the
possibility of such an effect obtained by the dipole moment method
is formed by Syrkin's results [79] showing that (1 : 1) complexes
of the nonpolar molecules of bromine and iodine with dioxane in

---

* Under this name we shall include organic molecular complexes of the donor-accep-
tor type. The nomenclature is extremely indefinite at the present time (see [77]).

TABLE 54. Dipole Moments of Complexes (1 : 1) of
Iodine with Some Amines and Sulfides in Benzene

Donor component	μ_{donor}, D	$\mu_{complex}$, D	Literature
Ethylamine	1.28	6.6	[80]
Diethylamine	1.03	6.2	[80]
Triethylamine	0.87	12.44	[81]
Diethyl sulfide	1.63	4.62	[82]
Dibutyl sulfide	1.64	5.00	[82]
Tetrahydrothiophene	1.73	5.49	[82]
Diphenyl sulfide	1.56	1.99	[82]

benzene* have dipole moments of about 1 D, and the complex of
iodine with pyridine is characterized by a moment of 4.17 D, while
pyridine itself possesses a moment of only 2.2 D. An even larger
increase in the dipole moment is found in the formation of complexes of iodine with aliphatic amines and sulfides (Table 54).

The considerable increase in the dipole moments of the complexes shown in Table 54 (apart from the complex with diphenyl
sulfide) cannot be explained either by classical polarization effects
[78] or by anomalies in the electronic and atomic polarizations of
the complexes [81]. The cause of this is a partial electronic transfer from the nonbonding n-orbital of the donor to a low antibonding
orbital of the acceptor (iodine). The degree of electron transfer
will obviously depend on the difference in the energy levels of the
orbitals mentioned and will increase with a rise in the energy level
(i.e., with a decrease in the ionization potential) of the donor orbital. In actual fact, the increase in the dipole moment of the complex is at a maximum when the nitrogen and sulfur atoms are attached to alkyl groups not conjugated with n-orbitals. In the case
of diphenyl sulfide, there is almost no increase in dipole moment
in the complex (see Table 54).

Among the data given in Table 54, attention is attracted by
the particularly high dipole moment of the complex of triethyl-
amine with iodine determined in dioxane (with complete charge
transfer in the equilibrium configuration, the moment may amount
to about 18 D [80]). The interpretation of this value must be ap-

---

* The dipole moment of the iodine-dioxane complex in cyclohexane is 3 D [83].

TABLE 55. Dipole Moments (in D) of Organic π-Complexes
in Carbon Tetrachloride [78]

Donor	Acceptor		
	tetracyano-ethylene	chloranil	2,4,6-trinitro-benzene
Durene	1.26	—	0.55
Hexamethylbenzene	1.35	1.0	0.87
Naphthalene	1.28	0.90	0.69
Stilbene	—	—	0.82

proached with care, since it is not excluded that the complexes of tertiary aliphatic amines in dioxane solutions exist in the form of ion pairs of the type $R_3\overset{+}{N}I \ I_3^-$, or other types [78, 80, 81].

It is not only such strong acceptors as iodine that are capable of forming molecular complexes with compounds the molecules of which contain unshared electron pairs. Sharpe and Walker [84, 85] have obtained evidence that even carbon tetrachloride interacts with amines, sulfides, ethers, and phosphenes by a type of formation of $n \rightarrow \sigma$-complexes. However, in this case the increase in the dipole moments of the complexes is only about 0.1–0.2 D, which shows the very small contribution of structures with charge transfer.

Compounds, the molecules of which do not contain unshared electron pairs but possess a branched π-system with a high-energy bonding π-orbital, can act as the donor components of complexes. A special review [86] containing detailed summaries of the moments of complexes the acceptor components of which are salts and carbonyls of the transition metals has been devoted to the dipole moments of such π-complexes.

Table 55 gives the dipole moments of a number of organic π-complexes of the sandwich type formed by nonpolar components.

Although the observed moments are small, they cannot be explained by the induction of a dipole moment by the polar groups of the acceptor [78] and indicate some charge transfer, i.e., a contribution of the ionic structure ψ_1 in formula (VI.6). At the same time, judging from the dipole moment, the transfer of charge is very small (not greater than 3-5% of the ionic structure).

A similar result has been obtained for some other types of π-complexes: complexes of naphthols and naphthylamines with m-dinitrobenzene [88] and even the complex of benzene with $AlBr_3$ (a dimer), the dipole moment of which, according to recent accurate measurements, does not exceed 1 D [89]. Thus, charge transfer cannot be the main cause of the stability of π-complexes and it is therefore due mainly to Van der Waals forces. This conclusion is in good agreement with recently developed ideas on the structure of π-complexes [77, 87].

4. Dipole Moments of Molecules in
Electronically Excited States*

In just the same way as the dipole moments in the ground electronic states of molecules $\vec{\mu}_g$, to a consideration of which the whole of the preceding material has been devoted, are characterized by the electronic configuration of the molecule in the ground state, the dipole moments in electronically excited states $\vec{\mu}_e$ give an idea of the electron distribution in the corresponding excited state of the molecule. A comparison of the values of $\vec{\mu}_g$ and $\vec{\mu}_e$ gives an idea of the direction of the change in the electron density on the individual atoms of the molecule on excitation, which is of interest for an understanding of the connection between the nature of charge migration and the redistribution of the energy of electronic excitation, for the determination of the probability of radiationless processes of the deactivation of the excited state and, consequently, for the capacity of a compound to luminesce, for determining the reactive center of an excited state of a molecule in a photochemical reaction, etc.

The experimental determination of dipole moments of molecules in excited states cannot be based on the usual methods of measuring molecular polarization (Chapter II) because of the extremely short lives of these states and, consequently, the negligi-

---

* M. I. Knyazhanskii participated in the writing of this section, and to him the authors express their thanks.

ble concentrations of excited molecules under ordinary conditions of excitation. The methods of determining $\overrightarrow{\mu_e}$ are based on the optical properties of molecules placed in various conditions of electrical interaction with the medium (solvent) or with an external electric field and are associated with the study of one of the following phenomena: 1) band shifts in absorption and luminescence spectra in various solvents and at various temperatures; 2) the luminescence and absorption of solutions in strong electric fields (polarization luminescence and electric dichroism); and 3) absorption spectra of compounds in the vapor phase in an electric field (Stark effect).

Determination of Dipole Moments in Excited States

from the Effects of Solvents on Absorption and

Luminescence Spectra

The main relations connecting the spectroscopic shifts and the basic parameters of the solvent and solute have been formulated by McRae and others [90-93]. These relations contain a series of parameters the determination of which presents great difficulty, and they cannot be used for direct evaluations of $\overrightarrow{\mu_e}$. However, with certain assumptions [91] it is possible to obtain the approximate formula

$$\Delta v_A = (AL_0 + B)\frac{n_D^2-1}{2n_D^2+1} + C\left(\frac{\varepsilon-1}{\varepsilon+2} - \frac{n_D^2-1}{2n_D^2+2}\right) \qquad \text{(VI.8)}$$

where Δv_A is the displacement of the maximum of the absorption band in a solvent with dielectric constant ε and a refractive index relative to the value n_D; A, B, C are constants characterizing the dissolved molecule; L_0 is a constant characterizing the solvent;

$$C = \frac{2}{hc_0a^3}(\overrightarrow{\mu_g} - \overrightarrow{\mu_e})\,\overrightarrow{\mu_g} \qquad \text{(VI.9)}$$

and a is the radius of the Onsager cavity (see Chapter II).

A convenient modification of relations (VI.8) and (VI.9) is formula (VI.10) which is suitable for calculating $\overrightarrow{\mu_e}$ from the measured shift in the absorption band in mixtures of two solvents possessing approximately the same values of n_D and substantially different values of ε [94]:

$$v_1 - v_2 = C\left(\frac{\varepsilon_1 - 1}{\varepsilon_1 + 2} - \frac{\varepsilon_2 - 1}{\varepsilon_2 + 2}\right) \tag{VI.10}$$

When a and $\vec{\mu_g}$ are known, it is possible to determine $\vec{\mu_e}$, calculating C from the slope of the straight line formed by a plot of equation (VI.10).

Formulas (VI.8) and (VI.9) have been used to calculate the dipole moments of excited states of a number of organic compounds [91, 94-96].

Lippert [97, 98] and Mataga et al. [99-101], on the basis of the theory expounded by Ooshika [92], have derived relations connecting the change in the dipole moments on passing into a lower excited singlet state with the shifts in the absorption fluorescence maxima in various solvents:

$$v_A - v_F = f\frac{(\vec{\mu_e} - \vec{\mu_g})^2}{a^3}$$

$$f = \frac{2}{hc_0}\left(\frac{\varepsilon - 1}{2\varepsilon + 1} - \frac{n_D^2 - 1}{2n_D^2 + 1}\right) \tag{VI.11}$$

where v_A and v_F are the frequencies of the corresponding absorption and fluorescence maxima, h is Planck's constant, and c_0 is the velocity of light. Relation (VI.11) corresponds to a linear dependence of the Stokes shifts ($v_A - v_F$) on f, from which the angular coefficient $(\vec{\mu_e} - \vec{\mu_g})^2 / a^3$ is calculated. If $\vec{\mu_g}, a,$* and the angle φ between $\vec{\mu_g}$, and $\vec{\mu_e}$, are known, $\vec{\mu_e}$. can be calculated. If the angle φ is unknown, it is possible only to evaluate the magnitude $|\vec{\Delta\mu}|$. It is frequently assumed that $\varphi = 0$ or $180°$ [97, 98]. Then

$$\mu_e = \mu_g \pm 0.010\sqrt{fa^3} \tag{VI.12}$$

Bilot and Kawski [102] have discussed the problem of the interaction of the dissolved molecule with the solvent independently of the preceding investigations. By considering the interaction as an excitation and making use of Onsager's model ideas, Bilot and Kawski, unlike the preceding authors, do not neglect the polarizability of the molecules in the final formulas but introduce the mean polarizability which is the same for the ground and excited states (α). In the overwhelming majority of cases, as the authors show,

---

* It is proposed to take $a = 0.8\rho$, where ρ is the molecular radius [97].

the relation $2\alpha/a^3 = 1$ is satisfied. In this case, Bilot and Kawski's equation assumes the following extremely simple form [103] for the absorption (A) and luminescence (F) spectra:

$$\Delta\nu_{A,\,F} = - m_{A,\,F} \left(\frac{2n^2+1}{n^2+2}\right) \left(\frac{\varepsilon-1}{\varepsilon+2} - \frac{n^2-1}{n^2+2}\right) - \frac{3m_2}{2} \cdot \frac{n^4-1}{(n^2+2)^2} \tag{VI.13}$$

for the Stokes shift:

$$\nu_A - \nu_F = m_1 \frac{2n^2+1}{n^2+2} \left(\frac{\varepsilon-1}{\varepsilon+2} - \frac{n^2-1}{n^2+2}\right) + \text{const} \tag{VI.14}$$

To determine the dipole moment of an excited molecule and the angle between the dipole moments in the ground and excited states it is sufficient to know two independent parameters m_1 and m_2 and to use the following working formulas:

$$\mu_e = \sqrt{\mu_g^2 + \frac{1}{2} m_2 h c_0 a^3}$$

$$\cos\varphi = \frac{1}{2\mu_g\mu_e} \left[(\mu_e^2+\mu_g^2) - \frac{m_1}{m_2}(\mu_e^2-\mu_g^2)\right] \tag{VI.15}$$

The parameter m_1 is defined by the slope of the straight line produced on plotting the graph of equation (VI.14) and m_2 is found most simply from a graphical comparison of the theoretically calculated spectral shifts according to equation (VI.13) with the shifts found experimentally. The required parameter m_2 (and also $m_{A,F}$) are selected in such a way that the points lie on a straight line intersecting the coordinate axes at an angle of 45°. The method given was used by Bilot and Kawski to determine μ_e and φ of a number of derivatives of stilbene, fluorene, some phthalimides, naphthols, naphthylamines, and other compounds [102]–[104]. An extremely rational method of determining dipole moments has been proposed by Bakhshiev [105-107]. Starting from ideas on universal interactions between the molecules of a solvent and a dissolved molecule, from general physical considerations, without using quantummechanical methods of calculation, and on the basis of Onsager's model the author obtained an expression for the shift in the electronic spectra as a function of n and ε of the solvent [105]. Subsequently, Bakhshiev [107], on the basis of more general theoretical considerations and taking some other factors into account, obtained the following expression for the case most frequently encountered in practice of solutions at room temperature:

$$hc\,\Delta v^s = c_2\left(\frac{2n^2+1}{n^2+2}\cdot\frac{\varepsilon-1}{\varepsilon+2} + p^s\,\frac{n^2-1}{n^2+2}\right) = c_2 f\,(\varepsilon,\,n) \qquad \text{(VI.16)}$$

$$hc\,\Delta v^{A-F} = \Delta c^{A-F}\left[\frac{2n^2+1}{n^2+2}\left(\frac{\varepsilon-1}{\varepsilon+2}-\frac{n^2-1}{n^2+2}\right) + p^{A-F}\,\frac{n^2-1}{n^2+2}\right] = \Delta c^{A-F}\varphi\,(\varepsilon,\,n)$$
$$\text{(VI.17)}$$

where $\Delta v^s = (\Delta v_A + \Delta v_F)/2$ and $\Delta v^{A-F} = v_A - v_F$ — the Stokes shift.

The dipole moment μ_e and the angle φ between $\vec{\mu_e}$ and $\vec{\mu_g}$ are determined with known values of μ_g and a from the following formulas:

$$\mu_e = \sqrt{\mu_g^2 - c_2 h c_0 a_3}$$
$$\cos\varphi = \frac{1}{2\mu_g\mu_e}\left[(\mu_g^2+\mu_e^2) - \frac{\Delta c^{A-F}}{2c_2}(\mu_g^2-\mu_e^2)\right] \qquad \text{(VI.18)}$$

In a study of symmetrical molecules with parallel $\vec{\mu_e}$ and $\vec{\mu_g}$ it is possible to determine the two dipole moments independently from the formulas

$$\mu_g = \sqrt{\frac{c_2 h c_0 a^3}{1-k^2}}\,; \qquad \mu_e = k\mu_g \qquad \text{(VI.19)}$$

where $\quad k = \pm\,\dfrac{1-\dfrac{\Delta c^{A-F}}{2c_2}}{1+\dfrac{\Delta c^{A-F}}{2c_2}}\quad$ (+ when $\varphi = 0$ and — when $\varphi = 180°$).

The terms c_2 and Δc^{A-F} are found from the angle of slope of the straight line, p^s and p^{A-F} either from the condition that the straight line passes through the point for the vapor (absence of shift) or (if there are no data for the vapor) from the condition of the observance of the best linearity of the relationship under study, and a is found from the structural model of the molecule. Using the method that he had developed, Bakhshiev studied both compounds for the molecules of which $\varphi = 0$ or $180°$ [105, 106] and compounds with $\varphi \neq 0$ or $180°$ [107]. It is just the latter case that offers the greatest interest, since here it is possible to determine by Bakhshiev's method the direction along which electron migration takes place when the molecule is excited.

Several attempts have been made [108–110] to use the basic relations described above to determine dipole moments from measurements of spectral shifts with changes in the temperature. These methods permit the use of the results of measurements of the shifts of absorption and luminescence spectra [109] or only those of the luminescence spectra [110] in a single solvent at room temperature and fairly low temperatures. This type of approach is extremely convenient in working with compounds characterized by low solubility in a wide range of solvents.

Finally, we must mention the work of Abe [111, 112], who has proposed a method for determining the dipole moment and polarizability α_e of the excited state of a molecule which is not based on Onsager's model and therefore does not require the evaluation of the Onsager radius a. Abe's method can be used to determine μ_e and α_e for any excited states, including nonfluorescent states (for example, the n,π^* state).

Determination of Dipole Moments in Excited States

from the Polarization of Luminescence in an Electric Field

The phenomena of the polarization of the fluorescence of molecules in liquid solutions in an electric field (electric fluorescence polarization) was discovered by Czekalla [126], who gave a phenomenological classical theory of the phenomenon, developed a method for determining the degree of polarization of the fluorescence of solutions, and proposed to use it to determine the dipole moments of excited states of fluorescing molecules in solutions. The essence of the effect is that the action of a strong electric field (E $\simeq$ 150 kV/cm) leads to the orientation of anisotropic molecules which, on excitation by ultraviolet light, emit partially polarized luminescence light with a very low degree of polarization, depending linearly on the square of the field strength:

$$P_E - P_0 = 0.1 \, \frac{3 \cos^2 \alpha - 1}{2} \left(\frac{\mu_e}{kT} \right)^2 \left(\frac{\varepsilon + 2}{3} \right)^2 E^2 \qquad \text{(VI.20)}$$

where P_E is the degree of polarization of the fluorescence of the molecules in a field of strength E in a solvent with dielectric constant ε, P_0 is the degree of polarization of the fluorescence of the same molecules in the same solvent without a field, and α is the angle between the vectors of the moment of the electronic transition and the dipole moment of the excited state, which can be determined either from additional considerations on the structure of the molecule or by using the results of measurements of electric dichroism (see below).

Formula (VI.20) can be used only if the dielectric relaxation time of the excited molecules in the given solvent τ_ε is substantially less than their mean life in the excited fluorescence state, $\tau_F : \tau_F / \tau_\varepsilon > 3$.

A more rigorous theory of the polarization of the luminescence in an electric field for any relationships between τ_ε and τ_F has been developed by Liptay [127], who has shown that it is also necessary to take into account the band shift occurring in an electric field as the result of the difference between μ_e and μ_g. Neglect of this additional factor leads to errors in μ_e estimated at 12-30% rel.

Determination of Dipole Moments in Excited States

from Measurements of Electric Dichroism

The phenomenon of electric dichroism was first observed and explained by Kuhn et al. [113]. It consists in the fact that a solution placed in a strong electric field exhibits anisotropic absorption in the region of the absorption bands characteristic for the dissolved molecules. Consequently, natural light passing through a solution will be partially polarized, and the intensity of the polarized light passing through the solution will depend on the angle between the directions of the polarization and the electric field. Electric dichroism is, of course, explained by the partial orientation of the dissolved molecules in a strong electric field. However, it was subsequently shown that at the same time there is a shift in the absorption band of the solution under investigation as a consequence of a change in the dipole moment of the molecules on excitation. The theory of the phenomenon taking into account the band shift was developed by Liptay and Czekalla [114] and by Labhart [115]. The method of determining the dipole moment of excited states based on a study of the electric dichroism of solutions as a function of the strength of the electric field and also the measuring procedure were proposed by Liptay, Czekalla, and Wick [114, 116, 126] and Labhart [117].

In the Czekalla method, the degree of polarization of natural light of frequency ν in an electric field of strength 20-90 kV/cm (where ν is in the region of the absorption of the solution) is studied. On considering the simplest case of the same direction of the dipole moments in both states and of the dipole moment of the electronic transition on absorption the following relation is satisfied:

$$P_E = -2.3\, D_\nu \left(0.1\, \frac{\mu_g^2}{k^2 T^2} + 0.46\, \frac{(\mu_e - \mu_g)\mu_g}{hc_0 kT} \cdot \frac{d\log\varepsilon_\nu}{d\nu} \right) \left(\frac{\varepsilon + 2}{3} \right)^2 E^2 \qquad \text{(VI.21)}$$

where P_E is the degree of polarization of the light with frequency ν passing through a solution with optical density D_ν, molar (decimal) absorption coefficient ε_ν, and dielectric constant ε, with field strength E. Thus, from the slope of the straight line $P_E = f(E^2)$ determined from experiment, D_ν, and (d log ε_ν)/d$_\nu$ (from the curve of the absorption spectra) it is possible, knowing μ_g, to calculate μ_e. In a more complex case, this method enables the angle between the vectors $\vec{\mu_g}$, and $\vec{\mu_e}$ and the dipole moment of the transition $\vec{M}$ to be determined. In the method of studying electric dichroism proposed by Labhart, the intensity of light of frequency ν polarized at a definite angle to the direction of the field and passing through the absorbing solution is determined. In this case, the relative change in the intensity of the light J passing through the solution without a field (through a comparison cell) and in the presence of a field E is also connected linearly with E^2, and in the simplest case ($\vec{\mu_e} \| \vec{\mu_g}$) the formula has the form

$$\frac{\Delta J}{J} = -\frac{5.29\, D_\nu}{3} \frac{(\mu_e - \mu_g)\, \mu_g}{hc_0 kT} \cdot \frac{d\log\varepsilon_\nu}{d\nu} \left(\frac{\varepsilon + 2}{3} \right)^2 E^2 \qquad \text{(VI.22)}$$

Hence, it is also possible to obtain μ_e, knowing μ_g and calculating (d log ε_ν)/d$_\nu$.

A combination of the measuring procedure proposed by Czekalla and by Labhart gives a convenient method for determining not only dipole moments but also (in more complex cases) the mutual orientation of the vectors of the dipole moments and the transition moment. A detailed review of all existing methods of observing electric dichroism and electric luminescence polarization has been given by Labhart [117, 118].

TABLE 56. Dipole Moments of Some Derivatives of Benzene, Biphenyl, and Stilbene in the Ground and First Singlet Excited States

Compound	μ_g, D [123]	μ_e, D				
		spectroscopic methods			fluores-cence polari-zation [121, 122]	electric dichro-ism
		Lippert [97]	Bilot and Kawsky [102, 103]	Bakhshiev [105, 107]		
p-Nitrosodimethyl-aniline	6.90	—	—	11.7	—	12
p-Nitrodimethyl-aniline	6.93	—	—	—	—	14.7
p-Dimethylamino-benzonitrile	6.55	23	15.7	—	11.1	—
p-Dimethylamino-β-cyanostyrene	6.75	14	—	—	—	14
p-Dimethylamino-β-nitrostyrene	7.70	18	—	—	17.8	18
4-Amino-4'-nitro-biphenyl	6.42	18	14.5	16.3	22.2	23.0
4-Dimethylamino-4'-nitrobiphenyl	6.93	—	—	—	23.1	24
2-Amino-7-nitro-fluorene	6.80	25	17.3	19.7	19.2	23.0
4-Amino-4'-cyano-biphenyl	6.0	—	—	—	15.3	16.5
4-Amino-4'-nitro-stilbene	6.83	—	—	—	23.4	22
4-Dimethylamino-4'-nitrostilbene	7.42	32	24.7	24.8	25.2	26.5
4-Dimethylamino-4'-cyanostilbene	7.10	29	21.5	21	21.2	20

Determination of the Dipole Moment in Excited States from the Stark Effect

Very recently, determinations of the dipole moments of formaldehyde and propynal in their n,π*-excited states have been carried out [119, 120] on the basis of measurements of the Stark effect in these states of the molecules. Here, to determine μ_e it is necessary to observe the splitting of the fine rotational structure in the electronic absorption spectra under the influence of an electric field.

In this case the selection of possible subjects of study is greatly limited since a resolved rotational structure is observed only in the electronic spectra of vapors of diatomic and the simplest polyatomic molecules.

Comparative Characteristics of the Various Methods
of Determining Dipole Moments of Excited States

It is natural that each of the methods described possesses its advantages and disadvantages. Thus the methods based on measurements of spectral shifts in solutions are extremely simple in experimental technique and calculation procedure but are very approximate because of the imperfection of the model ideas on the influence of the solvent on the dissolved molecule.

The methods of studying solutions in a strong electric field permit a more accurate evaluation of the dipole moments, but they are extremely complex in experimental technique and in the treatment of the experimental data. Preference must be given to the electric dichroism method, which possesses great universality and considerable accuracy. The Stark method is the least universal, but it is the only method for determining μ_e of molecules in vapors and is the most accurate. It must be noted that the methods described above give different values of μ_e since they measure different dipole moments of the excited compounds. Thus, the method of shifts in solvents gives a value of μ_e averaged with respect to the solvents, the electric fluorescence polarization method gives μ_e in the equilibrium fluorescence state in a nonpolar solvent, and the electric dichroism method gives that in the equilibrium Franck–Condon state. The Stark effect gives the value of μ_e directly in the molecule present in the vapor. It follows from what has been said above that the different methods supplement one another to different extents and this makes it desirable to use a combination of them.

Main Results of Determinations of Dipole Moments
of Molecules in Electronically Excited States

Table 56 gives the results of determinations of dipole moments of some derivatives of benzene, biphenyl, and stilbene obtained by the methods described above.

The data given show in the first place the closeness of the results obtained by different methods of determining μ_e. On the other hand, they permit interesting conclusions to be drawn concerning the electronic structure of excited states of molecules of organic compounds.

A comparison of the values of μ_e and μ_g (Table 56) shows the substantial polarization of the excited state of a molecule in comparison with the ground state. Thus, μ_e for 4-dimethylamino-4-nitrostilbene shows an almost 50% contribution of the bipolar structure (I) to the excited state [97] due to the first $\pi \rightarrow \pi^*$ transition.

An increase in polarity on the electronic excitation of a molecule is characteristic for almost all π, π^* excited states. A determination of the direction of the vector $\vec{\mu_e}$ always shows charge migration from the donor group to the acceptor group [97, 102, 105]. This conclusion is in good agreement with the results of numerous theoretical studies [118].

Interesting results have been obtained in a study of the dipole moments of excited states of alternant and nonalternant hydrocarbons. It has been found [111] that the naphthalene molecule is nonpolar in the $^1B_{1U}$- and $^1B_{2U}$- excited states, as in the ground state, which agrees with the predictions of the MO theory. Conversely, the azulene molecule retains a dipole moment in the excited states [96], but the direction of the vector $\vec{\mu_e}$ in the first two singlet excited states is opposite to the direction of $\vec{\mu_g}$. This shows a transfer of electrons from the five-membered nucleus to the seven-membered nucleus in the transition to the excited state.

Measurements of μ_e for the first excited state of charge transfer complexes have given decisive evidence in favor of a marked increase in the polarity of a complex in the excited state.

Thus, μ_e for the sandwich-type complex (II) averaged for a series of solvents is 14 ± 3 D [121], and shows almost complete charge transfer from hexamethylbenzene to tetrachlorophthalic anhydride on excitation (the calculated moment of the D^+A^- structure is 16 D). The value of μ_g is only 3.6 D. A marked increase in polarity in excited states has also been reported for other charge transfer complexes [107, 124].

If, however, the passage of a molecule into an excited state is connected with an $n \rightarrow \pi^*$ transition, the polarity of the excited state of the molecules decreases in comparison with that of the ground state. Thus, the magnitudes of μ_e of the n, π^* excited states of formaldehyde and propynal determined by the Stark method are, respectively, 1.56 and 0.7 D. The corresponding values of μ_g are 2.31 and 2.46 D. Similar relationships between the magnitudes of $\mu_e(n, \pi^*)$ and μ_g are characteristic for other carbonyl compounds and indicate that for them the transition into the n, π^* excited states is accompanied with the polarization of the carbonyl group in a direction opposite to that which occurs in the ground state.

Results of recent measurements of μ_e for n, π^* states of pyridazine (1.1 D) and pyrimidine (0.5 D) lead to a similar conclusion: for pyrimidine the direction of the $\vec{\mu_e}$ vector is opposite to that of the $\vec{\mu_g}$ vector [125].

The study of the dipole moments of excited states is, basically, only beginning. The main value of the results so obtained consists in the fact that at the present time the dipole moment of an excited state is almost the only experimental magnitude characterizing the electronic configuration of an excited state.

Literature Cited

1. V. N. Vasil'eva and E. N. Gur'yanova, Zh. Fiz. Khim., 28:1319 (1954).
2. E. N. Gur'yanova, I. I. Éitingon, M. S. Fel'dshtein, I. G. Chernomorskaya, and B. A. Dogadkin, Zh. Obshch. Khim., 31:3709 (1961).
3. C. W. N. Cumper, R. F. A. Ginman, D. G. Redford, and A. I. Vogel, J. Chem. Soc., 1963:1731.

4. Yu. I. Kitaev, S. A. Flegontov, and T. V. Troepol'skaya, Izv. Akad. Nauk SSSR, Ser. Khim., 2086 (1966).

5. E. Bergmann and A. Weizmann, Trans. Faraday Soc., 32:1318 (1936).

6. F. Gerson, F. Gäumann, and E. Heilbronner, Helv. Chim. Acta, 41:1481 (1958).

7. C. W. N. Cumper, G. B. Leton, and A. I. Vogel, J. Chem. Soc., 1965:2067.

8. V. I. Minkin, O. A. Osipov, and V. A. Kogan, Dokl. Akad. Nauk SSSR, 145:336 (1962).

9. V. I. Minkin, O. A. Osipov, V. A. Kogan, and R. R. Shagidullin, Zh. Fiz. Khim., 37:1492 (1963).

10. A. Albert and J. P. Phillips, J. Chem. Soc., 1956:1294.

11. D. G. Leis and B. C. Curran, J. Am. Chem. Soc., 67:79 (1945).

12. T. W. Campbell and B. F. Day, Chem. Rev., 48:299 (1951).

13. C. F. Ferrado, J. J. Draney, and M. Cefola, J. Am. Chem. Soc., 75:206 (1953).

14. V. I. Minkin, L. E. Nivorozhkin, and A. V. Knyazev, Khim. Geterotsikl. Soedin., 1966:409.

15. S. A. Giller, I. B. Mazheika, I. I. Grandberg, and L. I. Gorbacheva, Khim. Geterotsikl. Soedin., 1967:130.

16. N. D. Sokolov, Usp. Fiz. Nauk, 47:205 (1955).

17. C. A. Coulson, Hydrogen Bonding, Papers Symposium Ljubljana, London (1959), p. 339.

18. C. A. Coulson, Research, 10:149 (1957).

19. J. W. Smith, Sci. Progr., 52(205):97 (1964).

20. G. Pimentel and O. MacClennan, The Hydrogen Bond, W. H. Freeman, San Francisco (1960).

21. N. D. Sokolov, Dokl. Akad. Nauk SSSR, 58:611 (1947); Zh. Éksperim. Teoret. Fiz., 23:315 (1952).

22. J. R. Hulett, J. A. Pegg, and L. E. Sutton, J. Chem. Soc., 1955:3901.

23. G. Rataichak and L. Sobchik, Zh. Strukt. Khim., 6:262 (1965).

24. E. N. Gur'yanova, in: The Hydrogen Bond, Izd. Nauka, Moscow (1964), p. 281.

25. J. W. Smith, J. Chim. Phys., 61:126 (1964).

26. K. Bauge and J. W. Smith, J. Chem. Soc., 1966 A:616.

27. R. J. Bishop and L. E. Sutton, J. Chem. Soc., 1964:6100.

28. M. Davies and L. Sobczyk, J. Chem. Soc., 1962:3000.

29. L. Sobczyk and J. K. Syrkin, Roczn. Chem., 30:881, 893 (1956); 31:197, 204, 1245 (1957).

30. J. A. Geddes and C. A. Kraus, Trans. Faraday Soc., 32:585 (1936).

31. W. Otting, Ber., 89:2887 (1956).

32. O. A. Osipov, A. M. Simonov, V. I. Minkin, and A. D. Garnovskii, Dokl. Akad. Nauk SSSR, 137:1374 (1961).

33. V. I. Minkin, O. A. Osipov, A. D. Garnovskii, and A. M. Simonov, Zh. Fiz. Khim., 36:469 (1962).

34. A. E. Lutskii et al., Zh. Fiz. Khim., 23:361 (1949); 33:174, 331, 2017, 2135 (1959); 35:1938 (1961); Zh. Obshch. Khim., 29:2073 (1959); 30:4080 (1960); 33:2328 (1963).

35. Ya. K. Syrkin and V. Vasil'ev, Zh. Fiz. Khim., 15:254 (1941); Acta Physicochim. URSS, 14:414 (1941).

36. A. V. Few and J. W. Smith, J. Chem. Soc., 1949:753, 2663.

37. J. W. Smith, J. Chem. Soc., 1953:109.

38. J. W. Smith and S. Walshaw, J. Chem. Soc., 1957:3217, 5427; 1959:3784.

39. J. H. Richards and S. Walker, Trans. Faraday Soc., 57:399, 406, 418 (1961).

40. A. E. Lutskii and B. P. Kondratenko, Zh. Fiz. Khim., 33:2017 (1959).

41. J. H. Richards and S. Walker, Tetrahedron, 20:841 (1964).

42. V. I. Minkin, Yu. A. Zhdanov, I. D. Sadekov, and A. D. Garnovskii, Zh. Fiz. Khim., 40:657 (1966).

43. D. N. Shigorin, in: The Hydrogen Bond, Izd. Nauka (1964), p. 195.

44. B. Pullman and A. Pullman, Quantum Biochemistry, Wiley, New York (1963).

45. A. Szent-György, Bioenergetics, Academic Press, New York (1957).

46. A. Pullman and B. Pullman, in: Molecular Orbitals in Chemistry, Physics, and Biology, A Tribute to R. S. Mulliken, Academic Press, New York (1964), p. 547.

47. B. Eda and K. Ito, Bull. Chem. Soc. Japan, 29:524 (1956); 30:164 (1957).

48. V. I. Minkin, Yu. A. Zhdanov, A. D. Garnovskii, and I. D. Sadekov, Dokl. Akad. Nauk SSSR, 162:108 (1965).

49. V. I. Minkin, Yu. A. Zhdanov, I. D. Sadekov, and Yu. A. Ostroumov, Dokl. Akad. Nauk SSSR, 169:1095 (1966).

50. A. Chretien and P. Laureut, Compt. Rend., 195:792 (1932).

51. I. A. Sheka, Zh. Obshch. Khim., 27:848 (1957).

52. O. A. Osipov and M. A. Panina, Zh. Fiz. Khim., 32:2287 (1958).

53. A. V. Few and J. W. Smith, J. Chem. Soc., 1949:2781.

54. D. Cleverdon, G. B. Collins, and J. W. Smith, J. Chem. Soc., 1956:4499.

55. F. Rossotti and H. Rossotti, Determination of Stability Constants and Other Equilibrium Constants in Solutions, McGraw-Hill, New York (1961).

56. R. J. Bishop and L. E. Sutton, J. Chem. Soc., 1964:6100.

57. O. A. Osipov and I. K. Shelomov, Nauch. Doklady Vysshei Shkoly Khim. i Khim. Tekhnol., 1959:253.

58. I. K. Shelomov, I. A. Kozlov, and O. A. Osipov, Zh. Obshch. Khim., 36:1957 (1966).

59. E. N. Gur'yanova and I. P. Gol'dshtein, Zh. Obshch. Khim., 32:12 (1962).

60. J. Malecki, Acta Phys. Polon., 28:891 (1965).

61. C. F. Jumper and B. B. Howard, J. Phys. Chem., 70:588 (1966).

62. W. Kauzmann, Quantum Chemistry, Academic Press, New York (1957).

63. J. M. Lehn and G. Ourisson, Bull. Soc. Chim. France, 1963:1113.

64. C. Treiner, J. F. Skinner, and R. M. Fuoss, J. Phys. Chem., 68:3406 (1964).

65. M. E. Hobbs and W. W. Bates, J. Am. Chem. Soc., 74:746 (1952).

66. A. E. Lutskii, E. M. Obukhova, and B. P. Kondratenko, Zh. Obshch. Khim., 33:3023 (1963).

67. K. Hofmann, Imidazole and Its Derivatives (Part 6 of The Chemistry of Heterocyclic Compounds), Interscience, New York (1953).

68. W. Hückel, J. Datow, and E. Simmersbach, Z. Phys. Chem., 186A:129 (1940).

69. W. Hückel and W. Jahnentz, Ber., 74:652 (1941).

70. K. A. Jensen and A. Friediger, Kgl. Danske Vid. Selskab. Math.-Phys. Medd., 20:1 (1943).

71. O. A. Osipov, V. I. Minkin, A. M. Simonov, and A. D. Garnovskii, Zh. Fiz. Khim., 36:469, 1466 (1962).

72. G. Devoto, Gazz. Chim. Ital., 63:495 (1933).
73. C. M. Lee and W. D. Kumler, J. Am. Chem. Soc., 83:1593 (1961).
74. A. Huisgen and H. Waltz., Ber., 89:2616 (1956).
75. E. Fischer, J. Chem. Soc., 1955:1382.
76. R. S. Mulliken, J. Am. Chem. Soc., 74:811 (1952).
77. M. J. S. Dewar and C. C. Thompson, Tetrahedron, Suppl. 7, 97 (1966).
78. G. Briegleb, Elektronen-Donator-Acceptor Komplexe, Springer Verlag, Berlin (1961).
79. Ya. K. Syrkin and K. M. Anisimova, Dokl. Akad. Nauk SSSR, 59:1457 (1948).
80. S. Kobinata and S. Nagakura, J. Am. Chem. Soc., 88:3905 (1966).
81. K. Toyoda and W. B. Person, J. Am. Chem. Soc., 88:1629 (1966).
82. I. G. Arzamanova and E. N. Gur'yanova, Zh. Obshch. Khim., 33:3481 (1963).
83. G. Kortüm and H. Walz, Z. Elektrochem., 57:73 (1953).
84. S. Walker, in: Physical Methods in the Chemistry of Heterocyclic Compounds, ed. A. R. Katritzky, Academic Press, New York (1963).
85. A. N. Sharpe and S. Walker, J. Chem. Soc., 1964:2340.
86. A. D. Garnovskii, O. A. Osipov, and V. I. Minkin, Usp. Khim., 37.
87. R. S. Mulliken, J. Chem. Phys., 61:20 (1964).
88. A. E. Lutskii, V. V. Dorofeev, and V. P. Kondratenko, Zh. Obshch. Khim., 33:1969 (1963).
89. I. P. Romm and E. N. Gur'yanova, Zh. Obshch. Khim., 36:303 (1966).
90. N. Bayliss and E. McRae, J. Phys. Chem., 58:1002 (1954).
91. E. McRae, J. Phys. Chem., 61:562 (1957).
92. Y. Ooshika, J. Phys. Soc. Japan, 9:594 (1954).
93. S. Basu, Advan. Quantum Chem., 1:145 (1964).
94. M. Ito, K. Inuzuka, and S. Imanisi, J. Am. Chem. Soc., 82:1317 (1960).
95. K. Semba, Bull. Chem. Soc. Japan, 34:722 (1961).
96. W. Robertson, A. King, and O. Weigang, J. Chem. Phys., 35:464 (1961).
97. E. Lippert, Z. Naturforsch., 10a:541 (1955); Z. Elektrochem., 61:962 (1957); Z. Phys. Chem., N.F., 6:125 (1956).
98. E. Lippert, W. Lüder, and H. Boos, Advan. Molecular Spectroscopy, 1:442 (1962).
99. N. Mataga, Y. Kaifu, and M. Koizumi, Bull. Chem. Soc., Japan, 28:690 (1955); 29:465 (1956).
100. N. Mataga and Y. Torihashi, Bull. Chem. Soc. Japan, 36:356 (1963).
101. N. Mataga, Bull. Chem. Soc. Japan, 36:620, 654 (1963).
102. L. Bilot and A. Kawski, Z. Naturforsch., 17a:621 (1962); 18a:10, 256 (1963); Acta Phys. Polon., 22:289 (1962).
103. A. Kawski, Acta. Phys. Polon, 25:285 (1964); Bull. Acad. Polon. Sci., Ser. Sci. Math., Phys., Astr., 12:179 (1964).
104. A. Kawski and L. Bilot, Acta Phys. Polon., 26:41 (1964); Naturwiss., 51:82 (1964).
105. N. G. Bakhshiev, Optika i Spektroskopiya, 10:717 (1961); 12:473 (1962); 13:192 (1962); Ukr. Fiz. Zh., 7:920 (1962); Dokl. Akad. Nauk SSSR, 152:577 (1963).
106. B. S. Neporent, N. G. Bakhshiev, and Yu. T. Mazurenko, Elementary Photo Processes in Molecules, Izd. Nauka, Moscow (1966), p. 80.

107. N. G. Bakhshiev, Optika i Spektroskopiya, 16:821 (1964); 19:345, 535 (1965).

108. I. V. Piterskaya, Optika i Spektroskopiya, Collection of Papers III, Molecular Spectroscopy (1967), p. 18.

109. L. G. Pikulik and L. F. Gladchenko, Dokl. Akad. Nauk BelorussSSR, 8:641 (1964).

110. G. Lober, Ber. Bunsenges. Phys. Chem., 70:524 (1966).

111. T. Abe, Bull. Chem. Soc. Japan, 38:1314 (1965).

112. T. Abe, Y. Amako, T. Nishioka, and H. Azumi, Bull. Chem. Soc. Japan, 39:845, 936 (1966).

113. W. Kuhn, H. Dührkop, and H. Martin, Z. Phys. Chem., B45:121 (1939).

114. W. Liptay and J. Czekalla, Z. Naturforsch., 15a:1072 (1960); Z. Electrochem., 65:721 (1961).

115. H. Labhart, Helv. Chim. Acta, 44:447, 571 (1961).

116. J. Czekalla and G. Wick, Z. Elektrochem., 65:727 (1961).

117. H. Labhart, Chimia, 15:20 (1961); Tetrahedron, 19, Suppl. 2:223 (1963).

118. H. Labhart, Experientia, 22:65 (1966).

119. D. E. Freeman and W. Klemperer, J. Chem. Phys., 45:52 (1966).

120. D. E. Freeman, J. R. Lombardi, and W. Klemperer, J. Chem. Phys., 45:58 (1966).

121. J. Czekalla and K. Meyer, Z. Phys. Chem., N.F., 27:185 (1961).

122. J. Czekalla, W. Liptay, and K. Meyer, Ber. Bunsenges. Phys. Chem., 67:465 (1963).

123. O. A. Osipov and V. I. Minkin, Handbook on Dipole Moments, Izd. Vysshaya Shkola (1965).

124. E. Kosower, J. Am. Chem. Soc., 80:3261 (1958).

125. H. Baba, L. Goodman, and P. C. Valenti, J. Am. Chem. Soc., 88:5410 (1966).

126. J. Czekalla, Z. Elektrochem., 64:1221 (1960); Chimia, 15:26 (1961).

127. W. Liptay, Z. Naturforsch., 18a:705 (1963).

Index